SIMPSON

This work has been supported by the Radcliffe Institute for Advanced Study at Harvard University.

. . .

In addition, the publisher and the University of California Press Foundation gratefully acknowledge the generous support of the Simpson Imprint in Humanities.

. . .

The publisher also gratefully acknowledges the generous support of the Director's Circle of the University of California Press Foundation, whose members are:

Stephen and Melva Arditti
Margit Cotsen
Harriett and Richard E. Gold
R. Marilyn Lee and Harvey Schneider
Rowena and Marc Singer
Stevens Van Strum
Lynne Withey and Michael Hindus

The Ghosts of Gombe

The Ghosts of Gombe

A True Story of Love and Death in
an African Wilderness

Dale Peterson

UNIVERSITY OF CALIFORNIA PRESS

University of California Press, one of the most
distinguished university presses in the United States,
enriches lives around the world by advancing scholarship
in the humanities, social sciences, and natural sciences. Its
activities are supported by the UC Press Foundation and
by philanthropic contributions from individuals and
institutions. For more information, visit www.ucpress.edu.

University of California Press
Oakland, California

Library of Congress Cataloging-in-Publication Data

Names: Peterson, Dale, author.
Title: The ghosts of Gombe : a true story of love and
    death in an African wilderness / Dale Peterson.
Description: Oakland, California : University of
    California Press, [2018] | Includes index.
Identifi ers: LCCN 2017036838 (print) | LCCN 2017043621
    (ebook) | ISBN 9780520969964 (ebook) |
    ISBN 9780520297715 (cloth)
Subjects: LCSH: Davis, Ruth, –1969. | Primatologists—
    Tanzania—Gombe National Park—20th century. |
    Biological stations—Accidents—Tanzania—Gombe
    National Park. | Ghosts—Tanzania—Gombe National
    Park—20th century. | Accidents—Psychological
    aspects.
Classification: LCC QL31.D392 (ebook) | LCC QL31.D392
    P48 2018 (print) | DDC 590.73678—dc23 LC record
    available at https://lccn.loc.gov/2017036838

27 26 25 24 23 22 21 20 19 18
10 9 8 7 6 5 4 3 2 1

To the memory of Ruth Davis, Geza Teleki, and Carole Gale

For the chimpanzees of Gombe

Contents

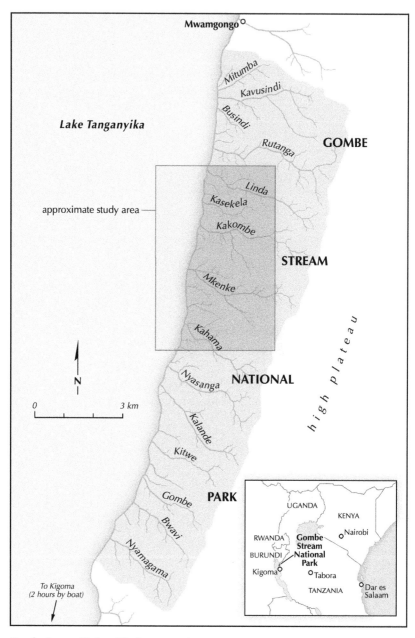

Gombe Stream National Park: an overview.

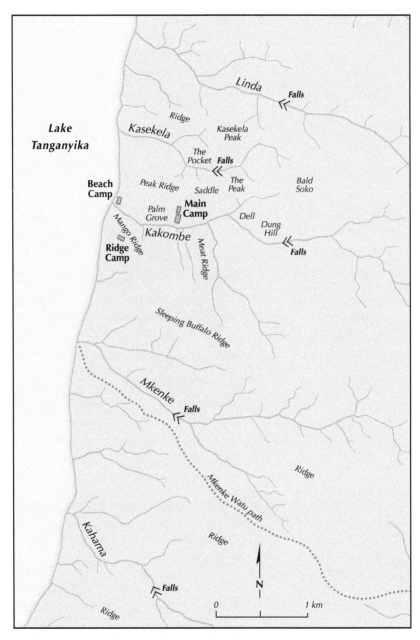

Gombe Stream Research Centre: the general study area.

Prologue

For a student of geology, flying in a small plane for the first time from Nairobi, Kenya, down to Kigoma, a town in western Tanzania at the edge of Lake Tanganyika, must be an exciting experience. Africa can seem like a mythical, magical place to any newcomer; but for someone interested in geology, that flight from Nairobi to Kigoma is a real-life classroom lesson at one of the most spectacular geological features in the world: the East African Rift.

The Rift! Where immense subterranean forces have slowly, during the last 25 million years, ripped open a deep gash in the crust of the earth. And when, on a clear day, you ride in that small, bouncing plane southwest from Nairobi to Kigoma, you may at some point be able to imagine reaching out and down to touch the rift, which from high enough above can look like a painful wound in the planetary skin. You'll discover as well, as you begin to move above the western edge of Tanzania, a place where that great wound has filled up with water, blue and glistening, and become Lake Tanganyika. You will also recognize that the high escarpment rising up so abruptly on the eastern shore of the lake is a stressed, compressed, and eroded edge of that giant piece of torn skin.

The airstrip at Kigoma is where you'll land, and you can walk or be driven down to the lake and take a boat north for a couple of hours until you disembark at one of the biologically richest and best-known forests in the world: the Gombe forest, which grows right there on the rift escarpment. It is an elongated rectangle of thick and tangled vegetation situated on the rift's moving edge, with one long side lapped by the waters of the lake while the other long side extends 725 meters up a series of rough and ragged reaches to meet the high Tanzanian plateau. Gombe's exceptionally rugged terrain was created by the flow of numerous streams that gather at that high plateau and descend, east to west, to the lake below—in the process, and over millions of years, scouring out numerous complex valleys. Fifteen of the streams are important enough to have been given names, which are also the names of the valleys. One of the southernmost streams is known as the Gombe Stream, which for unknown reasons has become the name of the larger ecosystem. And because the escarpment face exposes a few strata of especially hard rock, the streams tumble over cliffs and break into high waterfalls at a somewhat predictable point as they slide on their way down to the lake.

This special forest has long been protected from human intrusion not merely by its remoteness and terrain but also by cultural traditions and political decisions. The local Ha people, the Waha, may have regarded the Gombe forest as generally forbidden territory because it was said to include the sacred lairs of their formidable earth spirits. The Germans, who arrived in the late nineteenth century and claimed a good deal of East Africa as their own, formalized that early protection by defining boundaries and declaring the Gombe forest a special reserve for chimpanzees. With the collapse of the German colonial empire at the end of World War I, the British, working under a League of Nations mandate, took over the governance of Tanganyika Territory and, smartly following the German tradition, continued protecting that small rectangular forest at the western edge of their new territory, identifying it as the Gombe Stream Chimpanzee Reserve.

. . .

Tanganyika was still a British mandate and Gombe still described as a chimpanzee reserve when, in the summer of 1960, a twenty-six-year-old Englishwoman named Jane Goodall arrived. Accompanied by her mother, Vanne, and a recently hired African cook, Dominic Bandora, the young Miss Goodall pitched her tent in the forest not far from the shore of the lake and began her plan to study the chimpanzees. This improbable expedition was formally sponsored by the great paleoanthropologist Dr. Louis Leakey, but Jane Goodall had no scientific training. In fact, she had been Leakey's secretary, and she arrived at Gombe having no clear idea about how to go about studying the elusive apes. Neither, for that matter, had anyone else. No one had ever before observed wild chimpanzees to any degree, except perhaps for the one person who managed to publish a brief scientific account of them based on a few weeks of scattered observations done while crouching anxiously inside carefully constructed blinds. The precaution made good sense. As everyone knew, chimpanzees are immensely strong, emotionally volatile, and extremely dangerous.

Jane Goodall thought differently, though, and she tried a different approach. Instead of hiding, she let herself be seen. She moved through the forest without stealth. She showed herself openly, never acted fearful, and tried never to provoke fear. Gradually, and over several months, several of the apes became more or less bored with her nonthreatening presence, and gradually they began showing themselves to her. She was taming them, and in the process of doing so, she opened up for the first time in history the hidden world of wild chimpanzees. It was a remarkable accomplishment for which the young Englishwoman became world famous. Her fame happened quickly, assisted by a startling *National Geographic* magazine article that appeared in August 1963. Her reputation as a brilliant pioneering scientist developed at a slower pace, although she had already begun her studies at Cambridge University that would lead to a doctorate in 1965.

By the end of the decade, Dr. Jane Goodall was married to the Dutch wildlife photographer Baron Hugo van Lawick and had become the doting mother of an infant son. The ambitiously overextended mother and father were trying to care for their son, conduct field and photography studies in the East African savannas, write a number of scientific and popular books and articles, and continue the chimpanzee studies at Gombe. It was a challenging time, in short, made more so because their only significant funding source, the National Geographic Society, was starting to withdraw its support. In spite of a limited and unreliable budget, however, the van Lawick–Goodalls had already established a data-gathering routine at Gombe and were bringing in volunteers and students and young scientists to run the operation while striving to establish the connections and organization that would transform the original tented camp into a major scientific field station for animal behavior studies.

Then came a tragic event that marked the final summer of that promising first decade. At around noon, on Saturday, July 12, 1969, Ruth Davis, a young American working as a volunteer at the research site, walked out of camp to follow a chimpanzee into the forest and never returned. Six days later, her body was found floating in a pool at the base of a high waterfall.

Although Dr. Goodall was not in camp at the time, several young researchers and workers were. They were shocked and profoundly saddened by that event. Two of the researchers, Geza Teleki and Carole Gale, were devastated. Geza, who was Ruth's lover, happened to be in the United States at the time she disappeared, having been called back by his draft board for the physical examination preceding induction into the U.S. Army. Carole had been in camp on July 12 when Ruth disappeared, but she was feverish and bedridden from malaria and thus missed the first frantic days of the search for her missing colleague. At the end of the third day, however, Ruth appeared to Carole in a vision, one that felt entirely real, angrily demanding that Carole go find her. Carole obediently rose from her bed the next morning and, still weak from malaria, joined the search. Within three days, she had discovered the body, thereby accomplishing what well over a hundred others—the original search party having been expanded by dozens of regional police, camp and park staff, local fishermen, and area schoolchildren—had failed to do.

Both Geza and Carole were haunted by Ruth's death, their lives permanently altered. Both had shared with Ruth a quiet friendship and a secret perspective or insight, a gnosis perhaps—an extraordinary understanding of the chimpanzees, a special recognition that only the three of them shared, because only those three had spent many hours, days, and even weeks walking alone in the forest with those animals. In doing so, in moving out of the human world and into the forest world of the apes, they had begun to break through the psychological barrier that normally separates one species from another. They had developed strong personal attachments, genuine friendships, with individual chimpanzees; and in that way the chimpanzees became for the first time real to them, no longer faint spirits passing through from another dimension.

. . .

My own connection with this story began in 1989, twenty years after Ruth's death, when Jane Goodall was traveling regularly to Washington, DC, as part of a campaign—as one of the world's top experts on wild chimpanzees—to lobby members of Congress concerning the fate of chimps as determined by the U.S. Animal Welfare and Endangered Species Acts. She was profoundly concerned about the inappropriate and too often cruel and abusive treatment of chimpanzees in American biomedical research laboratories. She was also provoked by some clear indications of an international campaign within the research industry that would effectively circumvent recent legal protections limiting the destructive trade in live chimpanzees taken out of Africa. To assist her with her campaign and the political lobbying, Dr. Goodall had a colleague who lived in Washington and provided her with an office and pied-à-terre—his house—whenever she came to town. That colleague was Dr. Geza Teleki.

I was a writer hoping to write a book about chimpanzees and looking for experts to consult. Jane and Geza were experts hoping to create a book about chimps and looking for a writer to help. By the summer of 1989, the three of us had formed a partnership in order to produce a book that would detail some of the events and circumstances endangering wild chimps in Africa and expose the unethical treatment of captive chimps outside of Africa. I was to be the lead author and occasional investigator. Jane would be the second author. Geza would support us both with his own research and expertise and an enormous heap of documents stashed in a half dozen filing cabinets in his house. We had a good publisher lined up, and work was proceeding apace on a book that finally appeared in 1993 as *Visions of Caliban: On Chimpanzees and People.*

As for the distinctive name Geza Teleki, its national origin became clear to me after a casual look at objects scattered around his house in Washington. The old Hungarian paper money pressed under the glass top of the coffee table, for example. Or the sheet of antique Hungarian stamps framed and hung on a wall, each stamp a portrait of some aged gent in uniform looking inspired and wearing old-fashioned circular eyeglasses, the image repeated several times across and down à la Andy Warhol. Only later, after having made a number of trips down to Washington to confer with Geza—sitting across from him in his living room with the Hungarian-money coffee table between us, trying to figure out which of Geza's oddly divergent eyes to pay attention to, listening to the elaborate sentences as they rolled trippingly off his tongue, noting the occasional interruptions made by that bitter laugh of his—did I realize that the old man immortalized on the sheet of postage stamps was his grandfather.

Sometime after that I learned that his grandfather, Paul Teleki (or Pál Janos Ede Count Teleki de Szék), a Hungarian and Transylvanian count from a storied aristocratic family, was pictured on the stamps because he had twice been prime minister of Hungary. Eventually, I learned that Geza's father, Count Géza Teleki de Szék, had played field hockey in the 1936 Olympics and served as Hungarian minister of education during World War II. Ultimately, I discovered that Geza's prime minister grandfather had, at the start of that war, admitted to his most obvious foreign policy failure by shooting himself dead in the parliament building on the day the Nazi tanks rolled in. If history had gone one way, in other words, Geza might have become a dashing Hungarian and Transylvanian count, complete with castles and grand estates and all the rest. History went another way, and the family—young Geza, his father and mother—escaped in 1949 with little more than their own lives from the Soviet masters who had taken over Hungary.

That family background helped explain Geza's occasionally imperious manner, which could be off-putting. And yet, I thought, he and Jane together made a good pair. They were both charismatic individuals with strong personalities who seemed to balance and support one another. She had the understated style and social grace. He had the physical presence and political instinct. They seemed to make an effective yin and yang. I knew nothing at the time about the ghosts of Geza's past.

In any event, working on that book was the beginning of my education about chimpanzees. In America, I toured research labs and zoos and thereby discovered some tolerable situations and a few horrific ones. I met private owners of chimps and investigated the exploitation of apes for the entertainment industry in Hollywood and Las Vegas. I had traveled in Africa before and written earlier about African primates; researching this new book took me back to Africa for investigations and travel across eight nations in East, West, and Central Africa. I witnessed the breaking open of the great Congo Basin rain forest by men riding giant machines. I discovered young orphaned chimps held in crude cages or chained in junkyards and at doghouses. I saw the beginnings of a commercial trade in exotic meats for middle-class gourmets in urban Africa that would, within a few years, result in the slaughter of tens of thousands of chimpanzees and gorillas. And once Geza, in his role as guide and researcher and mentor, began pulling out his files and documents that identified the various excuses, abuses, cons, and conspiracies having to do with the human exploitation of chimpanzees for entertainment and biomedical research in the United States and parts of Europe, I began to appreciate the extent and complexity of what I was witnessing: the usually distressing and sometimes criminally perverse relationship between humans and their closest living relatives.

Those are some of the things Geza and I explored together during the writing of that book, and the shared knowledge and experience made us first colleagues, then friends. But I never felt that our friendship was an easy one. His laugh communicated irony rather than amusement. He had a grief and grievances to consider, most of which I only later began to understand. And yet we had traveled together, in an intellectual and emotional sense, and we had both stopped to drink at the same bitter stream. We had developed a common perspective, in short, a mutual understanding substantive enough that not long after his personal crisis came in September 2006, he thought to telephone me: "Dale, I need to talk." He wanted to talk about Ruth.

The conversation started slowly and continued, in intermittent bursts, for a few years, so that altogether I spent many days visiting him and his wife, Heather, in their new home in Bethesda, Maryland, listening to his story. Since we had agreed at the start that the story might become material for a book, our conversations turned into hours of tape-recorded interviews, and they led to other people and other interviews. Most especially, Geza introduced me to Carole Gale, who, as I noted earlier, had been a friend of his and Ruth's at Gombe in 1968 and 1969 and had been the one to discover Ruth's body. I came to like and admire Carole a great deal and to consider her a friend as well.

. . .

I have not yet described the state of Geza's health. Geza was dying during the period of our conversations, probably from having been poisoned. His thyroid was damaged. He had developed diabetes. His kidneys had failed, and so he was on dialysis for two years before his wife, Heather, gave him one of hers. He then got colon cancer. His immune system was compromised, and the antirejection drugs he took after the organ transplant accelerated his downward spiral, making him supremely vulnerable to every random bug. One of those delivered an end to his suffering in January 2015.

The poisoning was, Geza concluded, the consequence of many years spent living in suburban Washington, DC, in an older house situated downhill from American University. During World War I, American University had leased parts of its campus to the U.S. Army, which then created the second-largest chemical warfare laboratory and production operation in the world. Around twelve hundred army chemists there experimented with some of the most lethal substances known to humankind. They secured the substances inside munitions and containers next to tethered goats and dogs, and then they blew up the munitions and containers. When the air cleared, the researchers rushed out to see how quickly and in what gruesome ways the goats and dogs had died. They were producing hundreds of pounds of deadly toxins each day, sealing them inside artillery shells, mortar rounds, and glass jugs and storing them in large ceramic carboys. And in the end, after the war was over, they had a disposal problem, which the army solved in part by dropping their contaminated leftovers into deep pits on the college campus. Among the worst of the deadly compounds experimented on was Lewisite, an oily, colorless liquid smelling faintly of geraniums and based significantly on the better-known poison arsenic, which can persist indefinitely in the soil and groundwater.

That might clarify why all of Geza's son's pet mice died such strange deaths: popping up like roasted marshmallows, exploding slowly from the inside out. It could explain why their dog died young of thyroid disease. It may explain Geza's thyroid failure, followed by diabetes and kidney failure. He wrote several times to the army requesting that they test the soils on his property. At first they refused. When they finally came, they did so, he told me, in secret. He began appealing to the army to learn the results and also writing to the Environmental Protection Agency, then getting official notes of reassurance. Finally, the full report was released, and everything began to make sense. He went to see an expert at Johns Hopkins medical center, who told him to get out of his house. In 2003, the three of them—Geza, Heather, and their son, Aidan—moved from Washington to Bethesda, Maryland. That was too late for Geza. By the time he and I began the work that would lead to this book, he was dying a slow and painful death.

Carole Gale, too, was very ill when I interviewed her over several days near the end of August 2009. She was by then living with good friends in Oregon. She had been diabetic for many years, and so she was keeping track of her blood sugar and giving herself shots of insulin; but now she was also weak from cancer. She had decided to stop the standard medical treatments and instead was working to build up her body and support her immune system. She was eating raw food and getting nutrients and as many enzymes as possible by eating plenty of fruits and vegetables, as well as consuming Celtic sea salt and extra minerals.

Carole had planned to write her own book about Ruth and Gombe, and then Geza got in touch. Carole went to meet him in Bethesda and stayed at the house for a few days, and they talked about coauthoring a book. But then they recognized that, by themselves, they lacked the time and also perhaps the energy to write their story. As Geza had done, so Carole, too, spoke to me with passion and eloquence. She, too, was chastened by an awareness that her life was approaching its end.

I was moved by Carole's courage and Geza's anguish, and by the complicated story they felt compelled to tell, and I was honored to be a friend and serve as a confidant. We three were close contemporaries, and the developing narrative had, for me, a personal resonance: it seemed to be a generational tale, one with echoes of the war in Vietnam, social conflict and political assassinations at home, and the formative experience of reaching adulthood during the painful and exciting 1960s. The tragedy of someone dying so young and so suddenly, as Ruth did, is always wrenching, but I found I also strongly identified with Ruth, who began to come alive for me through photographs, correspondence, and the memories and writings of others.

As my reflections about Ruth and my conversations with Geza and Carole and others who were there continued to develop, though, I recognized that the story was carrying me beyond essentially personal interests and pushing me to explore the more complex knots of cultural conflict and social fragmentation, the dystopian dynamic that becomes possible when people from divergent backgrounds and orientations are forced to live in great intimacy and isolation simultaneously. Possibly, I imagined, *The Ghosts of Gombe* might best be regarded as a collective or cultural biography. But if so, I began to recognize, it would also be one anchored in science and the history of science—albeit a history from the bottom up: a portrait of life at a remote field station told largely from the perspective of the volunteers, students, and young scientists, many of them still endeavoring to make the challenging transition from eager amateur to seasoned professional while laboring to carry the science forward under difficult and sometimes dangerous circumstances.

And yet, of course, the central problem of this extended reconstruction of a time long ago and a place far away is much simpler and stranger than all that. It is the haunting, the loss Geza and Carole, I and many others still living have struggled with for too long now to understand. It is the death of a beautiful young person of promise in what should have been the best time of her life. How did it happen? Why?

I

The Visit

(September 27, 2006)

The address on the envelope was typed. From Raleigh, North Carolina. Postmarked with the date May 17, 2004. It had been sent to George Washington University, but by then he was no longer teaching there, so it was forwarded to his home. He tore it open, read it. The writer's name did not register at first. What puzzled him was that the writer had a Hungarian name. There was no connection to the name Davis, but she identified herself as the "family genealogist" for the extended Davis family, and she wanted to know what Geza could tell her about Ruth's death. More specifically, the writer wanted to know why Ruth "jumped over the cliff."

Why she jumped? The idea astonished him: to realize that this kind of thing was being talked about by God knows how many people. *Suicide?* Geza had never thought Ruth committed suicide. How wrong that was! How did this ridiculous rumor gain currency among her relatives? What pain the idea must have caused her parents! And yet the rumor of suicide, as he would later discover, led to other, sadder and more absurd fantasies. One was that Geza had jilted her, abandoned her in Africa, isolated and without the funds to leave. In addition, so one branch of the fantasy continued, she had been pregnant at the time. He had left her because of that. In desperation, she had taken a train across Tanzania all the way to Dar es Salaam to have a secret abortion before returning to the forest and the chimpanzees in order to jump off that cliff. It was a crudely melodramatic scenario, and it left Geza playing the role of villain. Other stories were in circulation, he eventually discovered, including the vaguely racist one that some "big black man" in the heat of nefarious pursuit pushed her off the cliff.

The letter was a shock. It really hurt. It was like being hit with something hard, only you're not really hit. It was like being in an earthquake. You don't see it. You hear it, but you hear it in your bones. When you're struck by a message like that, everything goes dead silent. Your ears just stop, and your mind shuts down. That's how it registered with Geza as he sat in his chair and read it for the first time.

The awful letter was folded up and put away. It got buried somewhere in the strata of papers on his desk, deliberately ignored. He couldn't handle it. He hoped it, and the memories and feelings and questions it had touched and opened, would flatten down and disappear. That did not happen. Instead, and little by little, the letter began to provoke him, to bother him. In part because he felt there was a good chance he wouldn't make it through another year, Geza decided to write a note in response. But that was all he would do. On September 22, 2006, nearly two and a half years after he had received the letter, he responded with a brief, impersonal email: "If you wish to pursue the matter, please respond, and I will do my best to clarify the rumors you mentioned."

Once he sent the email, however, Geza realized that he had to do more. It was one of those domino things. He didn't do it for the inquisitive genealogist. He did it because he could not bear the idea of those stories going on. He had been living in the shadowland of denial and amnesia, dropped into an oubliette of his own making, and now he began deliberately and methodically returning to those events of forty years ago: rummaging through old papers, examining old photographs and maps, writing to old friends and associates.

The vision or visitation or visit, or whatever you want to call it, happened during the night of September 27, five days after he had written that email and had begun to pry open the oubliette. He had been lying in bed, thinking about things. It's difficult to sleep with the kinds of physical malfunctions he had acquired. Sleep was not reliable. Not a condition with any longevity. He was lucky to average four hours of torn, scattered sleep a night. That night, he had drifted asleep in his usual restless, haphazard way, but when he woke again, something was different. Something was wrong. Fully awake now, he opened his eyes and knew he was not alone. Someone was in the room. It was the dead of night. A glow from the streetlamp outside slipped in between gaps in the blinds and faintly illuminated the room with a barred twilight.

Geza saw a human figure standing in the murk of the far wall. It was a woman with long dark hair, but her face remained shadowed and indistinct. He said, "What's wrong? What's wrong?" As she moved toward the bed, he understood through her style of movement (in the way you recognize a chimpanzee passing far away at the horizon by the manner in which he or she moves) who the figure was. It was Ruth. She came to the edge of the bed. Then reality split, and he watched as her hand pulled back the covers, while, at the same time, he saw the covers remain unmoved. She began to climb into bed, while, at the same time, nothing of the sort was happening. He shivered with fear or shock, scrambled to get out of the way, but the instant he did that the hand and moving figure vanished.

It was no ordinary dream. Geza had been fully awake and aware the minute he came out of sleep and sensed Ruth's presence in the room. He was awake the entire time, eyes open and seeing, and the experience was dreadful. Terrifying. He turned on the light, scratched down some notes about the event, and then lay awake for a long time.

They slept in separate rooms, Geza and Heather did, because he was such a restless sleeper. In the morning he went in to describe his experience, but before he had a chance, she told him she had been disturbed by a vision. *Vision* was the word she used, not *dream*. She had been waiting to fall asleep, passing into the relaxed twilight before sleep, when she saw a woman standing next to her bed, looking down and trying unsuccessfully to speak. Her face was pale, but the right side of her head was grievously injured. There was a dark, open crack on the right side of her head. When Heather sat up in astonishment, the vision faded. Neither Geza nor Heather had known that Ruth died with her skull crushed on the right side, nor had Heather ever seen a picture of Ruth. But when Geza showed her a photograph later that morning, Heather confirmed it: "Yes, that's who I saw."

A word to describe Geza's experience? Was it a vision? A visitation? A ghost? He didn't believe in any of that, although he was fascinated by the idea that people do feel as if they have seen such things. His father, who used to run séances in Budapest, was as much a scientist and a rationalist as anyone—and yet here he was calling out spirits. When it happened to Geza, he didn't have any feeling of contact with the supernatural. He had the feeling that an actual person was there, and that the person was Ruth. Period. He thought she was real, and he was upset. He tried to get away. He was afraid of her getting in the bed, which is what he thought was going on. True, he would not have been afraid of the real person doing that, but there was something extremely odd about it. He felt not fear so much as a sense of withdrawal that was purely involuntary. He wanted to shrink into himself. It wasn't a normal kind of fear, like in response to a dog coming at you. He thought it was real. There was no question in his mind that it was real.

But that was the first moment. After that, everything changed. After that, he began to realize that this could have been something he had dreamed up in his mind. It didn't bother him after that—until he went to Heather's room and spoke to her. That's when it really bothered him.

II

Beginnings

(November 1967 to June 1968)

1.

"Dr. Leakey?"

Leakey, white-haired, gray-mustached, full in the face, looked up from the mess on his desk and scowled. For a second Carole imagined herself through his eyes—this tall, young, eager American girl interrupting the great man's important paperwork, whatever it was—and she blurted out: "How can I get in touch with Jane Goodall? I want to work for her." She hadn't even introduced herself, and now she was feeling graceless.

Leakey: "I'll give you her phone number, and you can call her up." He scrawled a number on a scrap of paper, handed it to Carole, said, "Here. Call her."

Carole wasted nearly all her change on the museum pay phone, and it took an hour before she finally managed to reach Dr. Jane Goodall at the Grosvenor Hotel. But Dr. Goodall had a calm, melodious voice projecting an open personality, and Carole nervously explained that she was a third-year student at the Friends World Institute, an itinerant, Quaker-run experimental college from America that was based for the year in Nairobi. She wanted to volunteer for work—any kind of work, any kind whatsoever—having to do with Dr. Goodall's chimpanzee research, since she really loved animals. It was a friendly, polite, and positive conversation; and by the end of it, Carole had begun thinking of the person at the other end of the line as *Jane*. Not Dr. Goodall or Baroness van Lawick–Goodall. Baroness!

Jane had wanted to know if Carole could type and whether she liked babies. They needed a typist to support the chimp researchers at the reserve, as she called it. Occasional babysitting would be welcomed as well. Carole responded affirmatively to such queries and comments, and by the end of their talk, Jane had invited her to come visit them, her and her husband, Hugo van Lawick, at their home. She said, "Hugo and I always like to meet people before they go out to Gombe." Carole could spend a few days there, in fact, and since Carole had felt obliged to mention the interest of her roommate, Emma, in the same project, Emma was invited, too. Jane thought they could use a second typist at the reserve.

Carole spent another hour walking back to the FWI center, which was time enough to float in a billowing excitement snagged by a frustration that focused on her roommate, Emma. Or Em, as she was usually called. Carole had made the contact with Jane Goodall. Of course, the only reason Em hadn't was that Carole had, and only one person was necessary. Still, the frustration at having to share this glorious, life-changing opportunity with someone else, someone who was not passionate about animals in the way Carole was, took some time to dissipate.

. . .

Carole and Em spent a week with Jane and Hugo at their home in Limuru, which was several miles outside Nairobi. Jane and Hugo had begun renting the house in late February, only a few months earlier, whereupon she settled down long enough to deliver, on March 4, a baby boy named Hugo Eric Louis van Lawick. By the time Carole and Em arrived, in the first week of November, the baby was called Grub. That was an abbreviation for Grublin, a nickname the infant had acquired in the previous summer when, while the family stayed at the chimpanzee research camp, he displaced in reputation an infant chimp named Goblin Grub as the messiest eater in East Africa. The chimp returned to his original and simpler name, Goblin, while Hugo Eric Louis van Lawick became Grublin, then Grub.

Carole had heard about Jane Goodall and the Gombe chimpanzees before she went to Africa. She was on Christmas break during her first year in college, spending time with her second mother, who said something like: "Oh, we must watch television this evening. It's *Miss Goodall and the Wild Chimpanzees."* That was the first National Geographic television film about Jane Goodall, broadcast by CBS on Wednesday evening, December 22, 1965. Carole was, along with 20 million other American viewers, entranced by that shimmering vision of the brave young Englishwoman. Leggy. Blonde hair pulled into a ponytail. Smart. Brave. Graceful and understated. Living in a simple tent in an African forest with all those wild apes wandering in and out of camp.

As it turned out, Jane looked just like the young woman whose image Carole had seen on television in California not two years earlier. It was strange: meeting someone for the first time who was already familiar, like an old friend or a relative. The real person was glamorous enough, though. She was young (only a dozen years older than Carole, more or less), but also self-confident, good-looking, married to a sophisticated European aristocrat, and living an exciting life with animals in Africa. Carole could never convince herself that she was capable of being glamorous like that, since she was tall and big, with big legs and a sensitive personality. It was hard not to feel awkward in Jane's presence.

The two-story house at Limuru was built of stone and covered by a red-clay tile roof. It included a cat named Squink and two German shepherds named Jessica and Rusty. It was graced with a couple of vegetable gardens, an expansive front lawn with flowers, a rear stable large enough to hold four horses, and an eighty-mile view out the front windows. Jane and Hugo slept in a master bedroom upstairs. Carole and Em were given their own rooms at the first floor level: one in the recently built guest wing, the other behind the kitchen at the back of the house.

Her first morning there, Carole was surprised by a six o'clock knock on the door. Outside the door, she discovered a tray with hot tea in a cozied pot along with milk, sugar, and buttered biscuits, which was Carole's introduction to a tic of civilized Englishness. Jane's Englishness, in fact, produced not only tea twice daily but also her finely modulated speech and reserved manner, which at first made Carole feel boisterous in the American way, with a little naïveté thrown in for good measure. During a casual conversation that week, as Carole would later remember, she asked Jane what her favorite reading subjects were, and Jane said she had "very catholic interests." Carole, puzzled, said, "You mean religious?" Jane said, "Oh!" She laughed, then said, "No, it means just kind of widespread or universal."

Limuru was in the Kenyan highlands, which meant cool evenings. After dinner, they all warmed themselves in front of the fireplace, talking and drinking scotch. Hugo was a Dutchman by birth, a wildlife photographer by profession, and by inclination a raconteur who enjoyed holding forth on the adventures and dangers of life in the wild. In between Hugo's wonderful and often hilarious stories, they conversed generally about subjects of mutual interest, such as animals, conservation, tourists, and poaching. As Carole also learned during these conversations, Jane and Hugo, along with baby Grub, were now spending much of their time in the East African savannas—the Serengeti and Ngorongoro Crater in Tanzania—photographing and researching vultures and various carnivores. The vultures used stones as tools to break open ostrich eggs, which was Jane and Hugo's new discovery of animal tool-use that would be featured in an upcoming *National Geographic* article. The research on carnivores was for a book being financed by advance money from a British publishing house. That was Hugo's project. He had signed a contract to write it and take the photographs.

But they were a family, and all three—including Grub, of course—would be heading out to their work in the savannas, so Carole and Em would have to travel to the chimpanzee reserve on their own. Once they got there, they would not be alone. Jane had established a routine for the chimpanzee observations, and that routine was now managed by trained observers who were supported by a first-rate African staff. Jane and Hugo would drop in whenever they could. Fly down, most likely. They would also keep in touch by radiotelephone and then come to stay for the whole summer. There would be cooked meals, a private place to sleep and bathe, other Americans to keep them company, and enough typing of the daily scientific record to keep all their fingers and thumbs fully occupied. Jane wondered whether the girls were prepared for the isolation, but there was only one way to find out. She mentioned, also, that they should bring two of everything, including clothes and, since both Carole and Em admitted to being nearsighted, prescription glasses.

So long talks of romantic adventures and legendary places, of vultures and hyenas and chimpanzees, of impractical visions and practical necessities enlivened the evenings in Limuru. During the day, Hugo would drive off to Nairobi to shop and attend to other chores in preparation for the next photographic expedition, while Em and Carole visited with Jane and the baby and endeavored to make themselves both agreeable and useful.

One afternoon, Jane handed Carole the baby, saying, "Can you take him? I need to sleep." Carole took Grub out into the warm sunlight and bee-kissed flowers of the front yard, where he was fine for about two hours. Then he began to whimper. Carole tried to cheer him up. She bounced him, lifted him up and down. He laughed for a moment, cheered up, but started whimpering again, which was followed by crying. Carole thought Jane could hear them through an open bedroom window, and finally, when she could no longer do anything else, she took Grub back inside, carried him up the stairs, and knocked on Jane's door. Jane said, "Oh, bring him in. I think he just wants me. He's hungry."

That was that, but Carole also knew she was being evaluated, and she felt she had, in handling baby Grub that day, passed part of the test. Indeed, as Jane confirmed by the end of their visit, both Carole and Em were accepted. They would be given the chance to work as volunteer typists at the reserve during a three-month trial period.

. . .

On November 14, 1967, therefore, Carole and Em packed their bags—Carole pressing into the middle of hers the last of her marijuana stash, which was ten rolled joints—and were flown in a small plane south. They passed out of the big city and suburbs and moved over miles and miles of dry savanna, with hard waves of heat rising up unevenly from the flat, dry earth and making the flight bumpy, before at last dropping down at the town of Tabora, in Tanzania. They spent the night in a rundown hotel waiting for the next day's evening train headed west.

That train tumbled through the night and brought them, on the morning of November 16, to the end of the line, which was the town of Kigoma on the edge of Lake Tanganyika. They stepped into a daylight so fierce it made Carole reel, and they were greeted by two young white people, Americans both, who introduced themselves as Patti Moehlman and Tim Ransom. They were from the reserve. She had a Texas drawl and straight blond hair. He had a restrained manner and unrestrained dark hair and beard. Carole soon learned that he was from California, or at least had been a student at Berkeley, in California, and she liked his sweet smile, as well as the bushy dark beard and long hair, the beaded necklace he wore, and the beaded belt. Carole had seen that sort of beadwork in Nairobi, sold as a Maasai design, but perhaps it was better to imagine American Indians. It was a romantic style, in any case, and she understood it that way.

Kigoma was sleepy, dusty, and hot, with the train station down by the harbor and, running up from the harbor, a single paved street lined with small shops and one or two rough hotels on either side. Patti and Tim had planned their weekly supply trip to coincide with the arrival of the train, and there was still shopping to do. Once that was finished, they hauled everything—including a week's worth of food and other supplies—down to the harbor and climbed into the *Pink Lady,* which was a sixteen-and-a-half-foot white fiberglass Boston Whaler with a forty-horsepower outboard. Tim started the engine, and soon they were speeding out of the harbor and slicing a curve to the right and onto the open lake, heading north and following a scalloped shore.

Lake Tanganyika was as clear as glass yet soft, too, as if the glass were coolly molten. You could see right down into the water, far down, until the clarity turned into a pristine blue before slipping into a shadowy blue obscurity. Looking to the left, miles to the west on the far side of the lake, Carole could make out the distant green hills of the Congo, which rose and rolled back inland in waves, merging as they did into a faint haze and turning from green to blue to gray.

The sun pasted itself on bright and hot, but the heat was drawn away by the wind and a cooling spray of water. Sitting in the boat, Carole was not thinking so much as feeling: about how life was serious and how her three months in an African game reserve would draw her closer to the core purpose of her life. The nearby shore on her right was slashed with red erosion scars and marked by villages and rectangular patches of cultivation. After a time, the villages and cultivated patches stopped; and then the boat passed a narrow, jutting peninsula, a headland, and Patti shouted over the sound of the engine that they had just passed the southern boundary of the reserve. The brown hills turned to green, and instead of hot, barren, and sometimes eroding land, there was cool air and dark vegetation.

After another hour or so the engine was cut, and the boat crunched onto a pebble beach. Three Africans appeared and helped drag in the boat and unload the supplies. Then the four of them picked up their personal luggage; passed through a brief zone of fishy stink; and slipped into the shade of forest, a sizzle of insects, and the rank scent of moist earth and organic decay.

2.

The next morning began as a speckled rectangle of light seeping into a room. Carole, raised from sleep by rustlings and whisperings—and seeing Patti at the door—got up, threw on her clothes, grabbed a quick breakfast (toast with margarine and jam, instant Nescafé), then left the cabin to join Patti outside for the first shift. *Shift.* It sounded like factory work, an assembly line.

It was chilly and wet outside, drizzly and gray and not yet dawn, and Carole paused to take it all in: the tree-studded grassy meadow and two cabins, both made from bolted-together aluminum panels fixed on top of concrete slabs, the bigger one about fifteen by thirty feet, the smaller one the same width and half the length. Both cabins opened to the air and light through wire-mesh windows—no glass—and were protected from the weather by gabled roofs darkly insulated with a thick grass thatch held in place with chicken wire.

The bigger cabin, where Carole and Em had slept, was called Pan Palace. Aside from the large central workroom with chairs and two long tables with two typewriters on top, Pan Palace had a small bedroom to the left and another small room on the right that served as a minor kitchen where people could make coffee or tea and prepare their individual breakfasts in the morning. This kitchen contained a table, two chairs, a gas-bottle two-burner hotplate, and a kerosene-powered refrigerator.

The smaller cabin, called Lawick Lodge, consisted of a single room and served as the directors' office when Jane and Hugo were there, a storage depot for all duplicate records, and a retreat when the presence of too many chimps and baboons outside required it. Lawick Lodge also contained general supplies and a mimeograph machine for spitting out hand-drawn maps and other documents.

Then there were the boxes: about forty steel boxes embedded in the earth with concrete and scattered around on the ground like forty treasure chests, which was what they were, having been discreetly filled the previous night with chimp treasure. Bananas. The boxes were locked shut. The latches, Patti explained, could be opened remotely by pressing buttons on four panels inside Lawick Lodge. Battery-powered. Couple of car batteries. The juice running along buried wires. Very clever, very smart, but then you had to be smart to outsmart chimps. They went into Lawick Lodge, and Patti showed Carole how the banana boxes worked. Patti looked out through the open, mesh-covered window. She pressed a couple of buttons, and Carole heard the click of latches remotely withdrawn. The boxes were fixed in ways that allowed gravity to do the rest. Lids dropped open, exposing the treasure inside. It was a way, Patti explained, to keep the chimps coming into camp but still under control. No person was allowed to handle bananas in front of a chimp. That way, the chimps, who were far stronger than most people could imagine, wouldn't learn to associate bananas directly with people and, therefore, would not on a bad day kidnap someone.

The two of them went back outside into the continuing drizzle and waited: Carole standing next to the old steel oil drum on the front lip (the veranda, Patti called it) of the concrete slab for Pan Palace, Patti standing higher on the grassy slope, her tape recorder in a leather case strapped on her shoulder, microphone in hand.

A chimp appeared from out of the forest, the dark shape emerging from the deeper darkness and moving quickly and silently on all fours, galloping like a quiet horse into the clearing. More shapes emerged, appearing as if from another dimension in the oneiric shadows, from behind trees, from inside bushes and thickets, growing larger and becoming fuller and more real in the quiet, trickling morning. They were quick, and although Carole saw that she was taller than any of the chimps, especially when they were down on all fours, the chimps were still bulky and thick-limbed and, in truth, frightening. Especially the big males, who often, with raised hair on their arms and backs bulking them up like Olympic weight lifters, came racing, galloping, careening in, screaming and sometimes throwing rocks and branches. It was mostly display, Patti said. Typical male stuff you see anywhere, chimps or humans. Big ol' country boys coming into town and showing off, making a big fuss, announcing themselves, showing what big strong hulking macho men they were.

The chimps, Carole saw, were not beautiful, noble, or romantic. In fact, and being entirely honest with herself about it, she thought the chimps were ugly. They looked squashed and degraded. They had short squat legs. Arms long and muscular like a person's legs. Those huge and ridiculously dangling balls on the males! God! And the puffy pink pudenda of the females. Ugh! So big-mouthed, so explosive, so noisy: grunting, hooting, whimpering, screaming. Carole was thrilled to be there in a real African wilderness, but she really wished that the chimpanzees were lions.

The chimps had come into camp thinking *bananas!*—and it was Patti's job, speaking into the tape-recorder mic, to identify who was there and who was not, who was friendly with whom, who was good and who bad, who sick and who healthy, and to describe the comings and goings and interactions of everyone. Carole listened to Patti's steady commentary. She was like a baseball announcer on the radio, one with a mild Texas drawl giving a dutiful play-by-play description of the game: *Mike works up a pant-hoot, displays down north slope. He attacks Flo and Flint. Flo, screaming, carries Flint up a tree, and Mike runs back up the hill and beats on the oil drum. Worzle works up a pant-hoot, runs off*

into a tree west of camp. Sophie and Sorema arrive from northeast.
Leakey and Hugo come in from the east. Flo and Flint now in a tree
grooming. Pepe comes from the southwest, sees Mike, begins to pant-
hoot and shriek, gives a fear face, goes up and puts a hand on Mike's
shoulder, and Mike begins to groom him.

To Carole, they all looked the same, except for the obvious differences
between male and female, young and old. They were just big, scrambling,
hairy lumps. Hairy and scary. The males at least. The females, not so
much. Carole did her best to stay out of the way. She stood next to the oil
drum on the concrete slab, the veranda of Pan Palace, out of the line of
action. When Patti told her to, she would duck inside the door and watch
through the open window. Yes, sometimes they could be frightening,
those big apes, and after she had been outside for only two hours, she was
hit, accidentally to be sure, by a stone thrown randomly by one of the
males showing off. She wasn't hurt. And she was lucky that Patti, another
time, shoved her out of the way as one of the males came hurtling down
a path in her direction. It was dangerous out there, Carole saw, but she
also thought that the danger could be good. If you lived with it, adapted
to it, became part of it, then you wouldn't need to live in fear of it. You
would face it and begin to see that you're stronger because of it.

She also discovered, that first morning, another side to the chimps. A
male youngster—Patti called him Flint and said he was one of Flo's
children (whoever Flo was)—came right up to Carole and began hitting
her on the legs. He wanted to play, and then, after she didn't respond to
the invitation, Carole watched in amazement as little Flint began playing with one of the grizzled old males; together they tumbled and wres-
tled, both of them laughing as they raced about in a playful mock fight.
Laughing! Carole never knew chimps could laugh, but there it was: the
breathy, voiceless *ah-ah-ah-ah-ah* of a person laughing so hard he's
about to pee his pants. They didn't vocalize much, so it sounded like
someone sawing wood.

During that first day, Carole spent much of her time sitting inside
Pan Palace alongside Em, both of them in the workroom pounding at
the typewriters, transcribing the observation tapes, using both hands to
type and one foot to press a pedal that could stop and start the tape. But
she was still learning, and Patti kept her outside a lot of the time as well,
so that she could learn about the chimps, and when she did that she
hardly noticed the time passing. She was fascinated. Maybe twenty
chimps showed up that day. Patti said there were about thirty-five regular visitors, but they didn't all come every day.

Carole and Em type away inside Pan Palace, while Pepe, Charlie, and Hugo groom in the doorway.

A couple of chimps sat on the veranda and gazed at Carole, but mostly they concentrated on themselves and went about their business, which was grooming, displaying, hooting, attacking, retreating, reassuring, nursing babies, and—as the latches clicked and the boxes were strategically opened one by one—gorging on bananas. Patti did her best to introduce the chimps. Flo was the old lady, she said, and Mike was the alpha male. Charlie was the pugnacious one. Worzle was the grizzled one with whites in his eyes, like a person. And Olly was the female with the droopy lower lip.

Then there were the baboons, a whole troop of them, maybe fifty, hanging back in the trees and bushes, careful to keep safely away from the dangerous apes but watching and waiting for an opportunity to steal a banana or two or five. The baboons were a kind of monkey, and the young ones were cute and playful, often jumping and splashing in the swaying, cushiony trees the way little monkeys will do. But, all things considered, these were not your standard attractive monkeys. They spent a lot of time walking on the ground. They had monkey-style fingers and thumbs on their hands and feet, but doglike snouts. And the big males had white eyelids. It looked like a bizarre makeup job, as if someone had painted bright white stripes on their eyelids. Big? The adult males were twice as big as the adult females. They were the size of German shepherds, and they had canine teeth that looked like small daggers, maybe three or four times the size of a dog's canine teeth. The males would lazily close their eyes, and so the exposed eyelids would flash white, like bright semaphores. Then they would yawn, give a long, lingering, wide-open yawn, and instead of looking sleepy they looked vicious, showing off those daggers. The male baboons were another thing to watch out for, another danger, Carole realized. At the very least, she saw, the baboons were an enormous nuisance, upsetting the chimps, getting in the way. They were creeping, crafty, scrappy, snatching, opportunistic thieves.

. . .

People described the main camp—the little meadow in the big forest where the two aluminum cabins stood, where the banana provisioning and all the action took place—as being upstairs. To get downstairs, you walked down a long and twisting trail for a mile, more or less, until you reached the beach. Tim lived in a hut—tin roof, walls of stick and thatch—down at the beach. That was called beach camp.

Halfway along the twisting trail between downstairs and upstairs, between the beach camp below and the main camp above, was ridge camp, where a cooked dinner was served once the day ended. The African staff, meanwhile, lived with their wives and children in their own camp, which was like a small village, located above the beach and a short distance south of Tim's beach hut. One of their standard jobs was to prepare the evening meal for the researchers and haul it up the trail to the ridge camp. It was a nice place to gather for dinner, in fact: a natural clearing with a panoramic view of the lake.

When it was wet, the researchers ate dinner inside a small pavilion—concrete slab, four poles, corrugated tin on top—at the ridge camp. There was no furniture other than a couple of foam mattresses tossed onto the concrete floor, with a dim, unsteady light provided by a single hissing lantern. When it was dry, they ate dinner outside the pavilion on a concrete patio, sitting on the same mattresses in the light of the same lantern. Nothing fancy. You helped yourself. Found a convenient spot on a mattress. Relaxed. You might be visited quietly by one of the two catlike creatures, a civet and a genet, who hid trembling in the woods there. Emerging from the folds of darkness at dinner time, they expected a tithing from dinner and sometimes tolerated a pat on the head. But they were wild animals, of course, and always ready to bolt.

People chatted with each other. On Carole's first night, that meant: Tim the baboon man, Patti the chimp lady, Alice Sorem another chimp person with a long braid hung like a brown rope down her back, and Patrick McGinnis with short hair who was yet another chimp person. And Em, of course, who like Carole had just gotten there.

A cool breeze rose up from the lake. The sweat soaking Carole's shirt dried off. People talked about what they were doing or hoped to do, and because, from the ridge camp, they could look over and across the lake, they watched the sun turn red and settle behind the purple mountains of the Congo. The ridge they were perched on was bathed momentarily in a warm flash of red, and the lake before them turned indigo, then black, while the moon became a pale, distant eggshell floating over a glistening obsidian expanse.

Patti seemed to be a cheerful sort, and Carole liked her, but she also imagined that perhaps Patti's cheerful manner was not such a good thing. Patti seemed to get irritated more quickly than the others, and she controlled it under a false smile. Carole also thought maybe Patti liked awkward and self-effacing Em better than herself, but that, of course, was a matter of taste. The others—Alice Sorem, Pat McGinnis, Tim Ransom—seemed old-fashioned. Carole was nineteen, and they were all in their twenties. They also were more conventional and had done real academic work at legitimate colleges, rather than mostly nonacademic things at a nonlegitimate noncollege like the Friends World Institute. Or FWI, as she usually called it. Tim was a psychology graduate student from Berkeley. Alice and Patrick had been friends—had dated, briefly— and zoology students together at San Diego State. Alice also had done baboon work at the San Diego Zoo. Everyone had a college degree. They all had practical plans, schedules to enable the rational progress of their lives, and Carole did not. They were regular, normal people, she thought, open and friendly and without any visible neuroses. That, too, made Carole feel different. Such thoughts and observations flitted fitfully in her mind like the insects orbiting the lantern that lit up the last of the dinner, and they temporarily flew out of her mind when Tim the baboon man leaned over and said, in what seemed like an ironic manner: "And how was the first momentous day with the chimps?"

She answered him as directly as she could, but as she would later comment in her journal, her deeper reaction was irritation. She was irritated both by the question and the way it was asked. She wrote, *Why does everybody have to ridicule anything valuable and serious?* The irritation receded, and she regretted being so quick to judge, understanding that Tim did not really intend to ridicule her. He was just afraid of seeming serious. But then: *Why are people like that? Being serious is much more important than being silly all the time.*

3.

The rule was this: Any chimpanzee who came into the provisioning area would be identified and observed, and his or her activities would be recorded. All the records were typed, organized, and assembled into what people called General Records. Patti was the senior person in charge of observations and General Records, but she was going to leave in about a week, having finished her time at Gombe. After she left, Patrick McGinnis was next in line. He would be in charge. Then there was Alice Sorem, who had been at Gombe longer than either Pat or Patti, more than a year all told. But once you proved yourself for a year, you were allowed to move on to specialized research, to do your own project, which Alice was now engaged in. She was studying mothers with infants.

So Carole and Em were the two typists for Patti or Patrick, either of whom would be for the next several days in charge of observations— and always backed up by the other. On a long day there could be twelve hours' worth of chimps. Since the official observer couldn't leave during that time, Carole or Em had to fix food and carry it out to that person. Food that was uninteresting to the chimps, of course. Not apples or bananas. Scrambled eggs and toast, maybe, and cups of coffee.

The observer, Patti or Patrick, would finish a tape out there, hand it through the open window to Carole or Em inside Pan Palace, where they sat at the typewriters, and get a fresh tape. So they typed a lot, Carole and Em did, and they were also supposed to keep up the big charts that summarized the whole shebang: the operatic scenario of chimp society as it unfolded each day at the station. On the big charts, the chimps' names were abbreviated—HH for Hugh, HG for Hugo, and so on—and the nature of the behavior was color-coded. One color for aggression, another for play, and so on. But then, when the chimps had finally left for the day, and after the typing and charting were finally finished, they could all relax. Think. Read. Talk. Or catch up on the typing and records.

Soon enough Carole began to learn the routine, but she still was not resigned to having Em along on this big adventure, and the ambiguous nature of their relationship puzzled and rankled her. Maybe she knew Em too well, was too familiar with her habits and quirks: the unkempt hair, the bit lip, the evasive eyes. Maybe she saw Em as competition or an unhappy reminder of the embarrassing Quakers-in-a-Van school they both came from. And in spite of the reasonable voices coming from her better self, Carole was secretly hoping that Em, who was, after all, comparatively small and possibly fragile, would tire of the job and leave, at least by the end of their three-month trial period. Indeed, Em had already privately admitted to Carole that she was terrified of the big male chimps and couldn't bring herself to stand outside when the males were making their grand displays. So Carole considered that a good sign.

These were not thoughts to be proud of, and one afternoon, as the two of them sat in front of their typewriters inside Pan Palace, Carole decided to try breaking the shell of bad feelings by confessing everything. Perhaps the mortification of making such a confession would exorcise the feelings, and indeed, at the start of their conversation, Carole did feel relieved. But then the calm way in which her former roommate accepted all that anger while expressing earnest claims to understand and not be hurt by it soon made Carole more irritated. The lack of a real response to her petty feelings began to make them even worse, and so she became more emphatic in her assertions. Finally, it was late in the day and Patti suddenly appeared at Pan Palace, suggesting they all head down to the dinner hut at ridge camp. Em jumped at the suggestion, and her quick departure with Patti left Carole alone and feeling humiliated, since it showed that Em had gotten bored by the confession.

. . .

Patti had been living in a hut, an old rondavel down at ridge camp that was close to, but not quite visible from, the dinner pavilion. It was called a *rondavel* because it resembled a round hut of the South African style, but it was really octagonal and made of bolted-together aluminum pieces. Still, when Patti showed the hut to Carole after dinner that evening, Carole thought it looked sweet. It was big enough to hold a bed, desk, and chair. It had once been used for storage, and people said to watch out for scorpions, but when Carole saw it, she realized she wanted to sleep there. Patti said that as soon she was gone, Carole could move in.

Within the week, Patti was gone, and that's how Carole wound up sleeping alone in the middle of an African rain forest. It was a thrilling idea, although she stayed awake for a long time on her first night there, writing in her journal and then, after she had turned off the lantern, watching the final pinpricks of fading light dance and scatter, and being frightened by the well of darkness outside. She had grown up in a sub-urb of Pasadena, California, which was not a wild place, not many wild animals of any significant size, and yet she still knew she loved animals. Whatever the mysterious, original source of that attraction, it had been stimulated when she began reading J. R. R. Tolkien's *Lord of the Ring* series, which she loved because of the talking trees and the various creatures and people. She also read *King Solomon's Ring*, by Konrad Lorenz. That was a very different sort of book, of course, being nonfiction and science, but Lorenz was still a lot like Tolkien. The book made her feel that animals were not radically different from people, and that you could get to know them and even, possibly, communicate with them in a marvelous way. Had her early life been simpler, she might have followed that emotional interest in a more methodical and traditional fashion.

That first night spent alone in the forest, Carole found the leg-and-wing machinery of insects chirping and grinding and whirring to be a reassuring kind of background noise, but it would be interrupted by erratic bumps and thuds. More disturbing were the silences when the insect machinery paused for no obvious reason. And there were snakes to think about, a stirring tessellation of them weaving in and out of vis-ibility like the evil spirits they probably were: pythons sixteen feet long with folding teeth as big as dogs' teeth, night adders, burrowing adders, bush adders, puff adders, sand snakes, vine snakes, boomslangs, glossy black-and-white forest cobras, lemon-yellow spitting cobras, and six-foot-long black mambas with coffin-shaped heads. There was a lot to think about, and Carole felt afraid, although once she realized that she could bolt the door from the inside, and did so, she felt less afraid. The fear did not simply evaporate, however, and the next day, when she thought about it, she understood that there really were things to fear. She was afraid of the chimps and afraid of the forest, and yet she longed to go out into it, too. She wanted to be a wilderness person, but she was just a city kid from Pasadena, and now she had become a typist, spending most of each day typing up records inside Pan Palace.

. . .

Once Patti Moehlman left, Patrick McGinnis was in charge. Pat was on the conservative end of the spectrum. He was nice. He was uncomplicated, kind, and responsible, and Carole looked up to him. He knew what he was doing. But he was also conventional. You could see that in way he trimmed his hair and how he dressed. Even his body and facial hair seemed to follow rules, while his handwriting, which was microscopic, told Carole that he kept himself under a very tight regimen.

Meanwhile, Carole spent her time next to Em, the two of them inside Pan Palace and poking away: *tap-tap-tap-tap-tap-tap-tap-ding.* And yet all the action was happening outside. That's why when Patrick asked her if she knew how to work a camera, she said, "Yes!" She had taken pictures during her time at FWI, and she had also learned to develop her own photos. And now because of her supposed facility with a camera, Pat said he wanted her to take pictures of the females' pink bottoms, which included at the center their sexual parts, their vulvae, their pudenda. All the words for the subject seemed unpleasant, Carole thought, and of course there were much uglier words to consider, so maybe *bottom* was good enough.

Patrick soberly explained that the female chimps' bottoms changed shape and color in relation to their monthly cycles, with the greatest period of fertility being marked by the greatest swelling and brightest color. Maximum fertility meant maximum visibility, with the females' rear ends inflating until they resembled big pink pillows and functioned like big pink flags. The flags were being waved at the males, naturally, who all went a bit crazy in response. It was chimp pornography, but the problem for a scientific human watching the chimps and wanting to recognize fertility cycles was to know as precisely as possible which phase a female had reached. They were like phases of the moon. Was she fully swollen? Half swollen? Three-quarters? What did her fully deflated condition look like? The female chimps were individually about as different in that department as individual women were different in, say, hip or bust size. Patrick explained all that, and he said he wanted a photographic record, a simple way to document what the various stages of swelling looked like for each individual female. He handed Carole a Nikon with a split-image focus.

For her, it was an excuse to leave the typing for a while and go outside: following one of the females until she was about ten feet away from a pink bottom, then sighting and focusing and pressing the button. Carole did it with some distaste, imagining what it would be like if an alien from outer space went around taking pictures of human female bottoms, but she was young and glad to learn anything new, and it meant that she was outside with the chimps.

. . .

She was warned about the dangers. You had to show the chimps that they couldn't dominate you. People were mostly irrelevant to the chimps. They generally treated people like harmless bystanders who were not really there, were mostly invisible nonentities. But the chimps had this amazing strength, while people in comparison were absolute weaklings, just blobs of soft butter. When the chimps did notice people, they could be dangerous, and if they ever learned how weak people really were, realized how easily they could dominate a person, then everyone working there would be less safe. That's what Patrick said. So you did not want to give the impression of being afraid or weak. Patrick told her that the most dangerous chimp of all was old David Greybeard. Yes, he was Jane's favorite, and he was actually gentle. But he would wrap his steel fingers around someone's wrist or arm and not let go, thus taking that person hostage until a ransom in bananas was paid. Pat worried that one of the other chimps would figure out what a successful enterprise kidnapping was and start to do the same. But aside from David Greybeard, Patrick said, the only other ones Carole really needed to worry about were Hugh, Charlie, and Rix. Any of those three males would charge her and maybe slap out casually with a hand and knock her down.

"What do I do?" she asked. He said, "Don't run away if they charge. You have to stand still and face them." "But if they hit me?" "You jump sideways at the last minute, and then keep your feet and don't show them that you're afraid. Make sure they see that you can stand up to them."

So Carole knew about that. And during her first couple of weeks, she was periodically able to leave the typewriter and go outside with the camera, photographing pink bottoms. She saw no sign of any of the four males Patrick had warned her about, because they didn't appear often.

During that time, though, Carole got to know Mike, who did not charge people, even though he would come in making a grand entrance, displaying. There was that big empty oil barrel on the veranda, and Mike would scramble over, pounding across the ground with his feet and knuckled-down hands, maybe dragging a palm frond while galloping on all fours, the hair on his back and shoulders and arms ruffled up

Mike displays at the oil barrel in front of Pan Palace.

like turkey feathers as he reached the veranda. He would swing one of his great arms and slap that oil barrel with an open hand to make an enormous BOOM!!!! Or he would scramble over and hammer on the side of Pan Palace itself—BANG BANG BANG BANG!!!—and the metal building would shake. If it was early in the day and someone was still sleeping inside, it was like being inside a giant metal drum. A good wake-up alarm for the lazy. The late sleeper would drag on her shorts and shirt, get her shoes on, and go out. Patrick would already have started the observations. He was always ready.

Then one day Hugh showed up, and of course Carole didn't recognize him, since she had never seen him before. Patrick was standing on the upper part of the slope, looking over the meadow, and he saw past Carole to the southeast, where some big shadow had just emerged from the forest. He said, "That's Hugh coming." Just as Pat said that, Hugh spotted Carole, and his emotions possibly registered some message like: *There's a new human being! It's time to test!* He rose up from a crouch, stood up on his hind legs, and lifted his long arms. Stretched out like that, Hugh was the tallest chimp of the whole crew, and he ran at Carole with his arms up. They looked huge. With his arms up high, he seemed to be about as tall as she was, but he had those enormous muscles.

When she first saw him, Hugh was maybe thirty feet away but picking up speed. He was running directly at Carole. She could see his small black eyes, and they looked like shiny marbles in the head of some malevolent being. She knew she was supposed to stand still, but she just could not stand in the line of fire, so she started to creep sideways. He adjusted and kept running straight at her. There was no doubt he was aiming for her. Patrick yelled, "Don't move!" And so she stopped and waited for Hugh to get her. Then at the last minute, when she couldn't hold still any longer, when he was about to get her, she closed her eyes in terror, bent her knees, and jumped up and sideways. She had jumped two or three inches off the ground, but it felt like slow motion, as if she were in a dream and couldn't move fast enough. She had been sure he was going to get her, that his huge leathery hand was going to grab her face and maybe tear it off or grab her arm and throw her down. That didn't happen. She landed, opened her eyes, and he was gone. She was amazed. She turned and looked, and he was running up the hill. He had run right past her, and she had passed the test. She was amazed to have passed the test.

That kind of testing continued to happen over the next several days and then weeks, and Carole continued to pass the test. Each time she did, she gained more confidence. At the same time, once Patrick saw she could do it, he let her come outside more often. Em was smaller and slighter and not as athletic as Carole. Em could never get a good setup and a good jump. She was lightly hit once and that made her fearful, and then she ran. She had been intimidated. She began to come out mostly when the smaller and less aggressive chimps were present: the females or the males who didn't charge people. She preferred not to be out when the action took place. Carole was different. She used to play sports. She was a tall, strong tomboy, and she got a thrill out of being able to deal with the aggressive chimps.

4.

When Patti was still there, Carole had talked to her about trying to follow the chimps out in the forest, studying them away from the camp without the banana feeding. Patti had said, "That's what I'd do if I were going to stay."

Soon after her first month had elapsed, Carole asked Patrick: "Is there a chance that I could follow a chimp into the forest?"

He was discouraging. He said it wasn't possible. He said the chimps wouldn't like it. Trying to follow them would make you seem like a predator who was stalking them. They'd leave. "They'd just ditch you," he said.

Carole didn't know what to think, because she respected Pat, but she kept in mind what Patti had said before she left. Carole liked Patti and admired her courage.

On one of her days off, Carole decided to try it. The mothers caring for babies or dealing with juveniles were usually the slowest, since they had that extra burden, and Carole found a slow group: a mix of mothers, infants, and juveniles. She saw them wander out of camp, and then she followed them until they were just far enough out of camp that she could no longer hear any human voices. They stopped and sat down inside this wonderful little opening, a cave really, inside a thicket. The cave in the thicket reminded Carole of a childhood hideout. Not high enough for her to stand up in, but plenty of room to sit.

She sat down in there with them, and she relaxed completely, forgetting all the comments from Patrick about whether a person could follow the chimps anywhere. Then she realized how magical it was to be sitting there with those animals and to have them so comfortable with her, so trusting that even when she sneezed and made a sudden, awkward movement they weren't disturbed. They already knew her as a creature they'd seen, albeit faintly, in all kinds of situations, and now she was less faint. She was there. She was real and in their world. She thought, *This is wonderful, and it's so different from being in camp.* She really loved it: sitting with that small group in that wonderful hideout, the little secret forest place that was so peaceful.

Carole was by then accustomed to the sometimes frenetic pace of activity in camp. The chimps in camp had become used to her, and she was used to them, too. It was getting easy for her to deal with, say, one of the wildly displaying males. She had learned how to jump out of the way at the last minute, and she was learning to have eyes in the back of her head, to have that high level of alertness during high-pitched social events. It was thrilling. Of course, there were plenty of extended periods of quiet in camp as well, times when the chimps simply relaxed and socialized peacefully with each other, rather in the style of humans gathered at a picnic. But ultimately, what happened in camp was never completely real for Carole.

The chimps coming into camp didn't care about the people. The chimps came for the bananas; and when they got there, they played out their own social drama and then they left. And the people there—jabbering tersely into tape recorders, typing tappingly inside Pan Palace,

Charlie, Faben, Flo, and little Flint relaxing together.

sneaking furtively about in order to photograph females' private parts—were never partners in the chimpanzees' personal drama, not part of their lives. The people had tricked the chimps into coming into the clearing. Bribed them with bananas, and who could resist? The chimps played the game, but they didn't care about the creatures who had set it up. The people were faint and pale and mostly irrelevant, part of the scenery and close to invisible: dim figures perceived briefly behind a film of smoke, distorted reflections glimpsed fleetingly in a dark pool. The people were ghosts, actually, sometimes irritating ones to be sure, and sometimes a chimp had to put them in their place: show them who was who and what was what. But they were still ghosts.

You don't have sex with a ghost. You don't groom a ghost, not really. You don't seriously punch or pummel or punish a ghost when serious punching, pummeling, or punishment might otherwise be called for. You don't have an actual relationship with a ghost. Carole saw the truth of that, as strange as it might at first have seemed, and she saw that she had always been looking at the chimps from a different plane of perception and understanding. But now, sitting with them in that little bushy cave, she wasn't. Or so it seemed. And she didn't have to argue in her mind with Patrick about following them or not following them. She didn't have to worry about being ditched because *here she was. Here she was!* The chimps scratched itches on themselves, under their arms, behind their legs. The juveniles played. The mothers held their babies. At first they avoided looking at Carole, and then they didn't. They looked at her from time to time, then they looked away again. Mostly, they acted as if the situation were entirely normal: sitting in the cave with one of the strange ghosts from camp. For Carole, too, the experience felt normal, and wonderfully so. It really touched her, and she found herself thinking, *Wow, look at this! I get to be here in this wonderful world with these people I know.*

She had begun thinking of chimps as people. They were not people, of course, but they were so like people, and she was visiting their community, becoming part of it. The humans at Gombe were her community in one way, but in another way the chimps were her community, too.

. . .

On another of her days off, Carole followed Pooch. She was a young female who, at the time, had a terrible wound that had been plaguing her for a long while. It was a festering, open sore, and the wound made her less fit. She limped. So Carole thought Pooch would be slow enough to follow, and indeed she was. Pooch was stopping periodically to touch her wound and then sniff her fingers. She then would touch her genitals and sniff her fingers after that. She had not had a sexual swelling for as long as she had had the wound, and it seemed to Carole as if the two conditions were related, as if her body was trying to conserve energy and heal the wound.

After about three hours, though, Pooch just slipped away. Carole searched and searched, looking on the far side of Kasekela Stream, crawling through vines and undergrowth until she was scratched and bleeding. It hurt. She was also afraid, worried about putting her hand in the wrong place and touching, or stumbling across, snakes. But her long search for Pooch took her all the way to the Kasekela waterfall and the pool at its base; and when she saw where she was, she threw off her sweaty, dirty, torn clothes and splashed into the deepest part of the cold, limpid, wrinkling water. She dropped into the water and let the waterfall thunder around her ears. She floated naked and face down, dipping her hot, sweaty head under the surface and feeling the cold water reach into her hair and slide against her body. The coldness burned at first, and then it became cool, and she opened herself up to the experience of the water, which turned into feathers. She was feeling touched all over by the elements—the water and the air and sun—touched on her breasts, her thighs, being tickled and caressed, and she luxuriated in the feeling. She flipped over onto her back, laughed out loud, and let the current wash her downstream until she drifted into the shallows.

She stood up and allowed the sun to dry her off and warm her up, and she sat on a warm rock, leaned back, and took in the heat of the sun. She watched butterflies scatter above the banks, and she saw the sunlight skitter brilliantly on the water. As she would recall for her journal that evening, she felt completely and utterly at peace, wanting nothing more than the perfect paradise she had experienced at that moment.

. . .

Carole thought she did considerate things, such as always fixing breakfast for Patrick, from the thorn of guilt rather than the blossom of generosity. She did it because she wouldn't otherwise be able to enjoy her own breakfast. She would think Patrick was condemning her. Meanwhile, rather than feel angry at Em (who would get up late and fix her own breakfast at a leisurely pace, never offering to fix anything for Pat or Alice or Carole, all three of whom had been working since first light and were exhausted and starving), Carole found herself amazed. How could she do *that* without feeling crushed by guilt? Em's lack of guilt was impressive. Really, it was admirable. On the other hand, sometimes it was not.

They all depended upon each other to keep the work going, after all; and when, on one surprisingly clear day in January, Em took the best two hours of real warmth for her own swim, thereby leaving Carole as the single typist upstairs supporting Pat in the chimp observations, she was furious. When Em finally showed up to relieve her, Carole leapt up and began running down the path. She raced down, driving herself, killing herself, ran past the banana storage hut, through the palm tunnel, up the trail to ridge camp and finally up to the door of her hut. Out of breath but still angry, she tore off her sweaty clothes, yanked on her bathing suit, grabbed a mask and snorkel, then ran out and down to the lake.

Was it too late? The air was still warm. How was the water? She dove into the clear, clear water, felt the shock of cold turn into a caress of cool, kicked and kicked, and saw that she had entered the pale mansion of the underwater world. The surface wavered silver above while, below, the water gathered into sheets of blue and a hundred tiny fish flicked and

Tim and Bonnie at beach camp.

darted and joined tightly in a seeking cloud. She was alone, and she felt the water breaking into silver bubbles. She was remembering then forgetting about the people she wished she were closer to, forgetting them and forgetting the pain that people always seemed to bring. She swam parallel to the shore until, after twenty minutes or a half hour, she stopped and stood up, waist deep, to blow out her nose and clear the fog on the mask. The light had changed and an evening breeze had moved in, and it tingled and chilled her skin. She let the mask float on the water, and she recited out loud a few lines from *The Love Song of J. Alfred Prufrock* that had insinuated themselves in her mind. The recitation made her feel happy, and it cleared her mind. She walked out of the water and sat on the shore to let the water roll off, and then, when she was chilled, she ran back up to her hut and changed into dry clothes. After drying off and feeling clean and fresh, she walked back down to the beach to visit with Tim the baboon man and Bonnie, his new and newly arrived wife.

Carole thought Bonnie was lovely. She was short, sharp, opinionated, and she wore muumuus. She had a beautiful face that was sensual and warm. Her lips were full, her eyes brown, her skin tanned a deep brown. Her hair, which was a rich dark chestnut, fell into windblown wisps around her ears. Watching her in the fading light, Carole could see how beautiful Bonnie was.

They talked about Aldous Huxley, and Tim read a passage from Huxley's book *The Doors of Perception*. The passage went on and on about the need for a philosophical mystic to maintain perpetual vigilance against the ego. Carole didn't like Aldous Huxley, even though she considered herself a philosophical mystic, but she was careful to restrain herself, not to say anything negative, because she could see how much Huxley's words meant to Tim and Bonnie. And after Tim had finished reading, she left them in good spirits, then met up with them again at the ridge camp for dinner.

At the ridge camp dinner pavilion Carole was able to look out over the lake, look far off and see little cells of weather drifting here and there—storms, mostly, moving around like private floating tents with crazed activity inside: frantic twitches of lightning and furious debates between wind and water. Tim and Bonnie had brought up their tape recorder, so everyone listened to music that evening. Carole found the music poignant. When they played the tape of *Sergeant Pepper's Lonely Hearts Club Band,* she could only think of the previous summer when she was with her friends, the FWI kids in her group, driving around Switzerland in a blue VW bus for ten days, playing *Sergeant Pepper,* laughing and loving each other in their lonely way before separating at the end of it all to go on their different journeys. Thinking of those friends made her cry, and she hid the tears, letting her hair fall across her face in a curtain.

But when the music became cheerful and funny, she marked the beat with her fingers and sang the words, thinking how full her life was and how much like a character in a book she was. Somehow she always blamed herself for things—always believed, for example, that she was guilty of her own bad circulation and the unflattering ways in which her flesh adhered to her bones. She tortured herself into thinking she was a glutton because she had big legs. And yet there was no way, except starving herself and hiking ten miles every day (something she had tried and found impossible to maintain) that she could have thin legs. In spite of her body being not perfect, though, it was still a lovely body, and she had a handsome face. She was an individual, she thought, one of those young people who are filled with passion and living on the edge.

5.

Geza's first hours in Africa were bewildering. The bewilderment could have been partly the fault of his mother, who, laboring under the mistaken impression that her son had taken a sudden leave from his graduate studies in anthropology at Penn State University because he was headed to Africa to work for the famous Dr. Louis Leakey, wondered in a worried maternal way why Dr. Leakey had not informed her son in detail about the arrangements. She enlisted the help of a Teleki relative living in Europe, Lady Listowel, who phoned Dr. Leakey's office at the Centre for Prehistory and Paleontology in Nairobi two days before Geza was due to arrive and demanded to know what was going on.

Louis Leakey was still in the States at the time, so the long-distance call on Thursday, February 8, was handled by his secretary, Mrs. Crisp, who was already upset from having been swarmed by wild bees on Tuesday. That apiarian attack—twelve bee stings, seven to the head—had led to a 'peculiar feeling round my heart,' as the secretary soon wrote to her boss, and her condition was not improved by the call from Lady Listowel, whose sharp, superior tone had seemed simply rude. The voice on the phone had requested to know precisely *who* was going to meet Mr. Teleki at the airfield when his flight came in on Saturday. And *why* had no one reserved a hotel room for the young man?

Once she was informed about the impending arrival of this unrecognized individual, Mrs. Crisp did her job as she knew Dr. Leakey would have expected. She arranged for Mr. Teleki to stay at the Ainsworth Hotel for a week. Then she dashed off a letter explaining everything and had it delivered to the airport's customs officer with instructions to hand it over to the young American upon his arrival. Having done all that to the best of her abilities, Mrs. Crisp intended to rest over the weekend and recuperate from the bee attack. She had a "natural resilience" to such trauma, but the Teleki person's impending arrival was still an inconvenient and troubling mystery. "How all this muddle occurred," she wrote to Dr. Leakey, "and where Lady Listowel got the information that you had engaged him for 1 year, I don't know."

Meanwhile, Jane and Hugo's factotum in Nairobi, Mike Richmond, had asked his secretary to book a room for Geza in another hotel, the Devon; and, probably since Geza's flight was several hours late, neither Richmond nor anyone else met him at the airport. So Geza had no idea where to go or what to do after he stepped off the plane and into a heavy blanket of tropical air. Encumbered by two bulky suitcases and the normal disorientation of long-distance air travel, he spent several hours in the airport, smoking cigarettes and dropping Kenyan shillings into the pay phone while reviewing hotel listings in the phone book, hoping to find out which hotel he had been booked in. At around two o'clock in the morning, he learned it was the Devon. The taxi driver taking him into town spoke only Swahili. The night watchman at the hotel—grizzled, ancient, carrying a stick—also spoke only Swahili. They exchanged a few useless gesticulations in the lobby, after which Geza spied the board behind the check-in desk, retrieved a key, retreated to a room with a number on the door that matched the number on the key, and fell asleep with his clothes on.

The next day, Sunday, he met with some family friends in Nairobi and, in the evening, wrote a short aerogramme to his best friend back in Washington, DC—George Rabchevsky, a graduate student in the geology department headed by Geza's father—describing first impressions of Kenya and commenting on tourists, safaris, bikinis, and miniskirts. He promised to write again soon and ended with: "Please notify Ruth in case her letter got lost. Mails not too good here."

On Monday, he went shopping. He bought boots, a canteen, sleeping bag, plastic poncho, two sets of bush shirts and shorts, a rucksack, and twenty-eight rolls of film. The camera store owner was so happy with the big purchase that he had one of his assistants give Geza a ride back to his hotel. There, later that afternoon, a journalist phoned from the offices of the *East African Standard,* saying that he hoped to get an interview for an article, which he soon did. The article appeared a day later under the heading "A Second Teleki Explores E. Africa."

"Romantic escapades and noble sacrifices are a familiar part of Europe's aristocracy during this century," the piece began, noting that Geza's great-great-uncle, Count Samuel Teleki, had walked across East Africa during 1878 and 1888, becoming the first European to see Lakes Rudolf and Stephanie, which he named, honoring a couple of royal pals. He left his own name on the Teleki Valley of Mount Kenya and the Teleki Volcano farther north. The article then focused on the family's present scion, who had just arrived in Kenya and was now coping with time change and sunburn. "The story of Mr. Teleki's early life reads like a plot from an Eric Ambler thriller," the journalist wrote, "and he assured me that the reality is grim for someone who had to participate." The journalist went on to describe the dramatic family escape from Hungary after the war, their arrival and establishment in America, and young Geza's future plans. Geza's photograph, pasted in the upper-right-hand corner of the printed article, shows him seated and unsmiling, a bright sun placing his eyes in shadow and emphasizing the lean lines of a handsome face, aquiline nose, and dark mustache. He wears a dark sport coat and tie, and his thick, dark hair has been combed neatly back in a gleaming pompadour. He could pass for a young pop star from the late fifties or middle sixties, one of the Everly brothers, perhaps. But is that uncomfortable grimace caused by lingering fatigue and bright sun, is it a complacent glare, or is it the awkward expression of someone suddenly exposed as the untested descendant of a heroic figure from the past?

Geza had worked hard to grow up normal in America. He had learned to stop thinking in Hungarian, to master American English and baseball, and to become a maturing member of the postwar middle class. Following the example of his father, he refused to make much of the family's lost fortune, title, and status. That was ancestral baggage. The old stories were interesting, but in the bright and modern days of 1968—the promising new world where he was twenty-four years old and all grown up—little from the family's past was relevant. On the other hand, a title didn't mean much without the person behind it, and Geza's grandfather and father were intelligent, industrious, and bold men who had proved themselves in the blood sport of mid-twentieth-century middle-European politics. As for Count Samuel—the ancestor who walked through East Africa for thousands of miles while his contemporaries in Europe were busy dusting their wigs, and who survived the heat, rains, snakes, whirlwinds, wild animals, hunger, desolation, and cutthroat crosscurrents of intertribal wars and hostilities—he did so because he had the nerve, self-confidence, determination . . . the *balls* to do it. It took something along those lines to accomplish what he did. Geza could look back at some of his genetic predecessors and wonder if he had something to live up to. It was confusing, if you wasted time thinking about it: Do you ignore or admire your ancestors?

. . .

Early Thursday morning, he took a taxi out to Wilson Airfield, boarded a small, twin-engine plane that flew south, crossed the Rift Valley, and slipped over an outstretched knee of Mount Meru. The pilot then drilled a hole in the clouds and slipped into a miraculous netherworld: a two-thousand-foot-deep, one-hundred-square-mile sinkhole caused, two to three million years previously, by a cataclysmic volcanic explosion. It was Ngorongoro Crater, an ancient pock at the southern end of the Serengeti ecosystem.

Geza looked down and saw, far below, the flash of a pink-edged lake. A scattering of trees that were gathered into wrinkled, dark clumps. The dark vermiculation of a stream or two. A broad stretch of grassland occasionally slashed with stripes of pale brown and dotted by irregular spots and clumps of black. The spots and clumps turned out to be, as the plane spiraled lower, zebras and wildebeests, antelopes, buffalo, and God knew what else. The stripes of pale brown became artificial dirt tracks, of which one in particular, as they continued turning lower and lower, resolved itself into a crude airstrip interrupted by grazing zebras and wildebeests and marked by a flag at one end. The pilot passed low over the airstrip several times to drive away the zebras and wildebeests before dropping down to land.

Geza unbuckled himself and climbed out. The pilot helped unload Geza's two suitcases and one rucksack, and the pair of them stood there. It was quiet, nothing to replace the ringing roar of flight except the tick tick tick of cooling metal, the whistling of the wind, and the chomping chewing mooing sounds of grazing animals. Geza stood away from the plane to smoke a cigarette. The pilot smoked a cigarette. The wind blew. The animals ate. Then the pilot went through his checklist of wings, struts, tires, fuel levels, so on, and, after reassuring Geza that he would buzz over the van Lawick camp on the way out, he strapped himself in and pressed the button. The twin engines kicked in and started up. The props rattled and spun. The vibrating plane turned about, bounded down the dirt track, and took off.

A half hour passed, with Geza entirely by himself and standing next to his suitcases and rucksack, listening to the wind and the animals, smoking another cigarette. Then he saw a cloud of dust and heard the rumbling of a Land Rover that, as it hove into view, identified its master with bold letters painted on the side: BARON HUGO VAN LAWICK. *Why would anybody do that?* Geza asked himself. It was ridiculous, this European obsession with titles. And of course, he knew exactly what *baron* meant. The title happened to mark the lowest spot on the aristocratic totem pole. That was the last thing he would have painted on his own car, not that he would have painted anything else on it. The vehicle came to a stop, and a short, sunbaked, long-haired, unshaven, rather raggedy-looking man stepped out, shook Geza's hand warmly, and introduced himself as Hugo.

They drove on to the camp, where Geza was introduced to the Baroness Jane van Lawick–Goodall. Or Jane, as Hugo called her then, and as she clearly expected to be called. He was then introduced, in an abbreviated fashion, to Grub, an eleven-month-old baby with yellow hair, grubby face, and a few little white teeth starting to appear. Also in camp were Hugo's mother, Moeza, come to babysit; an artist friend of the van Lawicks named Bill, come to paint; Moro, a tall and pleasant-seeming African, who was the camp cook; Thomas, who was Moro's assistant; and Ben, a young American volunteer there to help Hugo with the cameras. Ben came from some kind of itinerant, international Quaker school temporarily headquartered in Nairobi, Hugo explained.

Jane was much as Geza had expected: gracious, barefoot, and slender, wearing khaki shorts and shirt, her blond hair clipped back into a ponytail. He hadn't imagined, however, that she would be so direct and uncomplicated, and he quickly concluded that they would get along. He liked her. He was also impressed by the camp, which was bigger and better-equipped than he had anticipated: with a wooden cabin and a stick-and-bamboo hut, six green canvas tents, and three vehicles (two of them VW buses belonging to Jane and to Bill the artist, the third being Hugo's Land Rover) parked to one side. Behind the camp gurgled a meandering stream ambitiously called the Munge River, and standing in the middle of the camp was an enormous fig tree, broad and thick enough to spread a pleasantly subaqueous light over the cabin, the hut, and the tents. The fig tree also hosted a troop of baboons, mostly invisible, who were eating ripe red figs.

The cabin had been modified for the baby's sake: protected outside against baboon incursions with a wrap of wire fencing, furnished inside with a playpen as well as a scattering of bright toys and a stack of clean diapers. There was also a bed for Jane and Hugo, plus a cupboard, shelves, and a grass mat cast over the stone floor. The tents had been divided between utilitarian and residential. One big tent Jane used as her office, with table, chairs, books and papers, a typewriter. Another big tent was the dining hall and sitting room. Four smaller tents were inhabited by Moeza, Ben, Bill, and now Geza. Bill had extended the veranda of his tent with a tarp, so he had an open place to paint. The two African workers, Moro and his assistant, Thomas, slept in the stick-and-bamboo hut, which also served as a kitchen, with the actual cooking done over an open fire outside.

It was a busy camp, with Jane, Hugo, and the baby often at the center of things. Geza joined Jane and Hugo for lunch in the shade of the open-walled dining tent, sitting back in a canvas chair and gazing lazily over a gorgeous vista that consisted of rolling grassland covered with animals, then a distant glittering lake followed by the rising wall of the crater rim. After a time, Jane retired to the work tent to resume typing, and Hugo, inviting Geza to come along, headed for the Land Rover.

. . .

They spent the better part of the afternoon passing and being passed by hundreds of animals: wildebeests and zebras, gazelles, elands, hyenas, jackals, a few elephants, and the occasional rhino, along with vultures and eagles and dozens of other birds of all kinds. Geza could barely absorb it all. Little more than a week ago he had been staying at his father's house in a chilled winter in the middle of Washington, DC, concrete and brick all around, cars whizzing back and forth, and now he was enjoying a hot summer within a walled kingdom of animals in the middle of Africa. It was astonishing!

Hugo sat across from him in the front of the Land Rover, regularly stopping to reach back into the open aluminum suitcase on the car floor behind them and pull out, from one of the baize-lined compartments, the right camera. Then he would find the appropriate lens and screw it onto the camera, screw the camera onto a camera mount fixed on the door, and take a picture or several. Both front doors of the car had camera mounts, in fact; and when the situation was good and the light just right, Hugo could drive—thoughtful slow or crazy fast, it didn't matter—while steering with his left hand, changing gears with the same hand, and holding a camera in his right. Hugo was good at multihanded behavior. It even seemed as if he could hold a camera, steer, shift, smoke a cigarette, and drink coffee simultaneously. Hugo smoked a lot, the old butts piling up and overflowing like a little waterfall from the ashtray, and in the same expression of ambitious excess, he gulped down his coffee, which was instant and spooned out of a can, sweetened with three spoonfuls of sugar.

Hugo had a coiled intensity that Geza soon began to appreciate, since he possessed some of the same nature. Hugo was also a talented and determined photographer, which was another thing Geza appreciated, since he considered himself a serious amateur photographer. Nothing like Hugo, of course. But Geza liked cameras and photography and had brought to Africa his favorite and only camera: an East German 35 mm, single-lens reflex camera with a 55 mm lens. He had bought it secondhand in Washington a year earlier and was eager to apply it to photons in Africa, but now he took pictures sparingly: partly because film and developing were so expensive and partly because he had the sensitivity to imagine that Hugo might not fully appreciate another camera-using person in the car.

Hugo was then concentrating his photographic efforts, he told Geza, on jackals and hyenas. He was doing a book. Photographs and text. He was not much of a writer, he confessed, especially in English, which was his second language, but maybe Jane would help with that. In any case, he wanted this book to be more than pretty pictures. The text would explain the animals, describe them as individuals with names and personalities and life histories, clarify how their societies worked. Fortunately, Hugo had just the previous month gotten out from under the thumb of the National Geographic Society, which for several years had kept him on a retainer. It had been steady work, but Hugo knew he could do better. The trouble with the Geographic Society was that when you worked for them, they owned you. They claimed rights to anything that came out of your camera, which for a professional photographer like Hugo amounted to being owned himself. It was unpleasant, restraining; and since Jane was the *National Geographic* magazine cover girl, his own subordinate role—*merely* the photographer, *only* the husband— was galling. It was hard to be always in your wife's shadow.

So now, while it was true that he no longer had a steady income, Hugo was free to be his own man. And he still had, for as long as it lasted, a decent advance against royalties given him last summer by his British publisher. The book was going to be called *Innocent Killers,* and Hugo hoped it would be a big success. The phrase *innocent killers* meant wild predatory animals—East African carnivores—who killed for their food. They didn't kill for fun or sport, as people did, so they were in that way "innocent." The book would feature six carnivores, Hugo imagined—golden jackals and hyenas first, then wild dogs, lions, leopards, and cheetahs. He had already done the photography for a couple of *Geographic* articles on hyenas and cheetahs. He couldn't use those photographs, but at least now he knew something about hyenas and lions, which was a good start.

Toward the end of the day, Hugo set up his system for night photography. He fixed to the outside of the car door a board with three flash bulbs powered by a battery and synchronized through electric wires to the shutter of his camera. He attached the wires, flicked a switch that connected to the battery. All he had to do after that, he explained, was find the action, spin the car around in the right direction so that the light would be aimed just so, wait for the precise instant when the action peaked—predator leaping to prey, for example—then press the camera button and, *poof,* a flash would peel open the dark world to reveal its bright white essence.

It was the wildebeest birthing season, and the hyenas would be out. Hugo hoped to photograph a kill. After the sun and moon traded places, therefore, he and Geza harnessed themselves in, strapped on crash helmets, and drove around without headlights under a creamy round moon, looking for hyenas until they sighted a racing pack of them, ghostly and giggling in their strange hyena way and bounding through the tall grass like a slavering pack of big dogs. Hugo took off, racing right behind the pack, banging and bouncing and swerving across the grasslands at thirty-five miles an hour. But the hyenas gave up the chase after a while, as did, eventually, Hugo and Geza.

. . .

As the days and then weeks of his introductory visit to Ngorongoro passed, Geza eventually came to appreciate above all Hugo's practical competence. If you were lost at night, stuck in the mud, and threatened by elephants, Hugo would find and extract you. If you were in danger from a pride of lions, he would drive them away, even if it took a day's effort. He knew what to do, and everyone else in camp would have been helpless without him. True, the *Baron* painted on the Land Rover was silly, but Geza came to see in him a capable, talented, determined man, a crack photographer who worked hard to photograph animals and who cared about them. From that perspective, then, the episode with Princess Margaret was unfortunate.

It happened late one afternoon while people were relaxing in the dining tent. A pack of Land Rovers raced up, stopped, and people got out. Geza didn't know who they were at first, but the van Lawicks must have had some warning. Jane, Hugo, and Moeza went over to the Land Rovers immediately, leaving Geza, Bill, and Ben still sitting in the tent. Geza saw an excited gathering next to the cars, and it soon became clear that Princess Margaret had arrived, along with some friends or family members. There were also secret service personnel in the front and back cars of the entourage.

Geza, Bill, and Ben were visible from the cars, and Jane waved to them and said, "Come on over." They went over. Hugo was just then introducing himself to Princess Margaret, saying, with a deferential bow, "How do you do? I'm Baron van Lawick." After that, Geza stepped forward and said to Princess Margaret, "How do you do? I'm Count Teleki."

It was something he had wanted to do for ages, particularly now that he was in East Africa where a previous Count Teleki's explorations were still well-known. It had seemed like a wonderful opportunity to introduce himself that way, to try out for the first time in his life that perspective of himself, and he did it in a positive spirit, thinking that Hugo would be amused. Hugo was not. His jaw dropped, his eyes narrowed, and for at least a week after that he hardly spoke a word to Geza.

6.

Carole and Em had both volunteered for three months. Em left early that February, while Carole settled in for a much longer stay, having typed out a long letter to the director of the Friends World Institute, passionately explaining herself and requesting that her three-month volunteer project be extended to at least a year. The director wrote back quickly, giving her official permission to continue at Gombe for the year, an extension that was soon seconded by Jane and Hugo. Carole was delighted by such developments, and thus, with her hopes raised so high for the future, she must have felt doubly betrayed by the illness that suddenly struck.

Yes, the Gombe mosquitoes have a lot to answer for. So does Gombe's hothouse environment, especially during the rainy months, when, as the atmosphere lowers itself down like a heavy lid on a simmering pot, warm combines with wet to create a nourishing soup for pathogens, including those that, after Carole began scratching at her mosquito bites back in early December, found a congenial home in the integumental disruption.

The bites became infected, in other words. Patrick gave Carole penicillin in ointment and tablet form. No effect. Pat then said she should go into Kigoma to the Baptist mission there and see Mrs. Owens, the missionary wife of the missionary preacher. She was from Tennessee and trained as a nurse. So one day in late December, Pat took Carole in the boat to Kigoma on the weekly supply run, and she was examined by Mrs. Owens, who said, as Carole would recall the words some forty years later, "Oh, this has become septic. It's in your bloodstream, so putting ointment on it and taking tablets is not going to do the job. You need an injection of penicillin, and I can give that to you." She added, "You wouldn't be safe taking an injection from the Egyptian doctor in the town clinic, because they don't have any sterilizing equipment, and they share their needles. But I have my own syringe, and I boil it up on my own kitchen stove every time." Carole liked and trusted Mrs. Owens. She had a big steel-and-glass syringe. She boiled it up and gave Carole an injection of penicillin.

By the end of January, Carole was back at Gombe and pleased to learn that she had graduated from mere typist to general researcher. She was good at the chimpanzee observations, and Patrick could use her help. Em, who hadn't left yet, was then the only full-time typist, while Carole and Patrick took turns outside, speaking their notes about chimpanzee behavior into tape recorders. Then Em left for good, going back to FWI in Nairobi, and Patrick and Carole began spelling each other with the observations and typing up their own notes afterward—until, on February 9, she became seriously ill.

She was still living in the ridge camp hut, and she walked up to the main camp early that morning as usual, ate breakfast, and started her observations. About halfway through the morning, she felt strange. She had no idea what was happening, but she asked for a break. Patrick took over, and Carole went into the kitchen of Pan Palace where there was a big plastic bowl. She leaned over and vomited into the bowl.

Too sick to do any more observations that day, she headed down to her hut and tucked herself into bed. During the night, she felt feverish, her skin burning, and then she felt cold and began shivering. A storm appeared, with explosions of thunder and searing flashes of lightning, and the bright flashes lit up the inside of her hut and etched the table, chair, and bed into her brain as perfect silhouettes, while she lay naked and shivering under her blanket and beneath her mosquito net. She got up, paced around, stood at the door, wanting, but also unwilling, to go outside into the rain; she watched, in another bright white flash, the palm trees outside being whipped by the wind. She slipped back under the mosquito net and the blanket, then sank into a lonely night of strange dreams and memories torn from childhood.

She remembered her mother. Her parents got divorced when she was very young, and then her mother was diagnosed with cancer when Carole was twelve and died when she was fourteen. That was her first major sorrow. She had another family—friends who felt like family—the Danas. Martha Dana was a dear friend of Carole's mother, and when she was in the hospital for two months around Christmas time with her first cancer surgery, Martha came to visit. Martha, talking to Carole's mother, learned that Carole did not enjoy being with her father, so she invited Carole to spend Christmas with the Dana family. Carole discovered that it was a wonderful family, very different from her own, with both a mother and a father and four happy daughters. They sang Christmas carols, and the environment was warm, friendly, and full of laughter. It seemed like the ideal family, and she began to think of the Danas as her second family and Martha Dana as her second mother. This idea took on a greater significance after her first mother one day said, "You know, Carole, I've got cancer, and it's possible I could die from this. Where would you want to live? Do you want to live with your father?" Carole said, "No, I don't want to. I want to live with Martha Dana and her family."

The Danas agreed to take her, but her father fought that plan in court after her mother's death, and so she was forced to live with him. Under those circumstances, her grades plummeted, which meant that when it came time to apply to college she didn't get in anywhere except some obscure New Hampshire college and that experimental Quaker college called the Friends World Institute, where everyone spent their time traveling around and learning from experience rather than from books. Life was funny that way, with the bad so unaccountably mixed up with the good.

. . .

Carole stayed in bed for a week: miserable, feverish, throwing up all the time. She reached a point where she could not even keep water down. Within a half hour of drinking water, she just threw that up as well. Patrick was by then giving her quinine, thinking she had malaria, but that wasn't it. Nobody who had malaria at Gombe threw up the way Carole did.

Finally, he took her in the boat back to the Baptist mission in Kigoma. Mrs. Owens was good as a nurse, and Carole liked her and also liked being mothered. She gave Carole sips of chicken broth. Carole had not been able to hold anything down, and Mrs. Owens tried to calm her stomach, give her just enough so that she could start retaining fluids. After two weeks she was eating chicken noodle soup and saltine crackers, and she had a small lettuce salad. She felt comforted by the chicken noodle soup and saltines because they reminded her of being home and being taken care of as a child.

Mrs. Owens gradually began to give her various bits of bland, solid food, and one day gave her a piece of doughnut, saying something like: "I fried this in my own oil. It's very good. Totally fresh. See how you like it." Although she never openly speculated about the possible connection between her patient's current condition and the general degree of sterility in that old steel-and-glass syringe in her kitchen, Mrs. Owens had begun to suspect that Carole might have liver problems caused by hepatitis. She had been giving foods that wouldn't stress Carole's liver, but she wondered if the doughnut would, since it had been fried in fat. Sure enough, as soon as Carole tasted that fresh donut, which was delicious, she threw up violently. It was the first time she had been sick like that in a week, and it led Mrs. Owens to conclude that her patient probably had hepatitis.

Based on that conclusion, Carole bought a ticket at the Kigoma station and boarded the next train to Dar es Salaam. By February 23 she had flown from Dar to Nairobi, where she went to see a British doctor, who looked at her and said, "I think you have yellow in your eyes. Can you spend a penny?" *Spend a penny* was a quaint British euphemism for *pee,* which was a euphemism for a blunter word that meant "urinate," which she did. After the doctor had tested her urine, he told her she had hepatitis and would have to avoid all fat and alcohol for at least six months. "And," he added, "you should rest the whole time."

She pleaded with him: "No, I want to go back to work. I just got the best job in the world." He said, "I think that's a bad idea. I think you should rest." She said, "But I feel better." She was nineteen years old and had just gotten a job in the best place in the world with Jane Goodall, and now she was going to quit? No way!

. . .

Geza's sojourn with the van Lawicks at Ngorongoro Crater lasted longer than anyone had planned. One night the lazy little creek known as the Munge River turned into a torrent that threatened to sweep away the entire camp, tents and all. They moved everything through knee-deep water to a dry spot on higher ground. But the long rains had come early that year, and the rushing water soon eradicated the two dirt roads into and out of the crater and made the landing strip inaccessible. On March 10, however, a break in the weather enabled a plane from Nairobi to swoop down and pick up Geza and Moeza, Hugo's mother, and carry them both to Kigoma. At Kigoma, the pilot traded Geza and his luggage for Patrick McGinnis and his, and then he flew both Patrick and Moeza to Nairobi.

Pat was planning a vacation. Moeza planned to fly from Nairobi back to her home in Holland. The thunderstorm that pounced so ferociously on their little aircraft as it pulled its way north along the edge of Lake Tanganyika challenged those plans, and for a time it challenged their sense of a stable reality altogether. The rushing water being chopped into pieces by their propeller threw a heavy gray shroud right over the windscreen, and it was impossible to see where they were going. Pat observed with growing concern that the pilot was tense and sweating. They made it to Nairobi, however, and upon stepping onto the tarmac at Wilson Airfield, all three headed straight for the bar. Didn't even think to pick up their luggage. Time for a drink.

Back at Kigoma, meanwhile, Geza stepped into the supply boat and was taken north on the lake, then led up a winding trail to the main camp and shown his room in the aluminum building called Pan Palace. The next day, he went to work on general observations. Or at least on training to do them. Since Patrick was gone, Alice Sorem set aside time from her own research project on mothers and infants in order to train Geza. Carole had just returned from her medical leave and was, she hoped, properly rested and repaired, so she, too, helped out with the general observations, trading places now and then with Alice and Geza. Carole's first impression of Geza? She saw a tall, lean, dark-haired young man in bush shorts and khaki shirt who acted sure of himself. He had a small mustache beneath a beaky nose, and she did not feel immediately glad to meet him. She reserved judgment. But she maintained in any case the notion that anyone who came to Gombe to study animals was going to be interesting almost by definition.

. . .

In the final week of March, Carole experienced a relapse. She was tired, couldn't do much, and so she went to bed, which meant, in turn, that Geza was suddenly given a lot of responsibility. Alice still helped him, made sure he was identifying the chimps and their behaviors appropriately, but he was increasingly responsible for the observations and typing, not to mention keeping up the charts and even taking care of other camp duties, such as financial records, supplies, and general maintenance. Perhaps that heightened responsibility accelerated his learning, but it was also stressful, and the stress may have exacerbated a tendency to look for subterranean motives in others. He began to suspect Carole of malingering. Why did she do nothing but lie in bed all day? Carole was a strong-looking young woman. She had lost about fifteen pounds from the hepatitis, but she came to Gombe with a few pounds more than necessary, so now she looked normal. She did not look like a sick person who was wasting away.

Geza mentioned his concerns to Alice, and she set him straight, saying something like: "The liver is a vital organ. You really are underestimating the problem." Bonnie, who overhead the conversation, relayed it back to Carole, and Carole appreciated Alice's sticking up for her. She did not appreciate the circumstances that prompted the sticking-up-for, though, and she remained uncertain about Geza. He had a fragile male ego, she decided, and could sometimes be a thoughtless prick. Geza, in turn, was not won over by Carole. She was flaky. She was a hippie. When not in bed being sick, she spent her evenings down at the beach with Tim and Bonnie smoking marijuana, listening to rock and roll, and reading out loud from a tattered volume of Shakespeare's plays.

The marijuana smoking was no secret, but Geza still disapproved. It was inappropriate at a scientific research site, and even if people hadn't been trying to do scientific research, he would not have approved. He did not believe in taking drugs for recreation. He thought Tim and Bonnie were fine, however, aside from that one bad habit. At least Tim was fine, given that he was, in spite of superficial appearances to the contrary, working to become a serious scientist.

Tim had gotten an undergraduate degree in psychology back East at Williams College in 1965. Then he went to Berkeley to do graduate work in psychology. His advisor had a special interest in hormones and behavior, and Tim spent his first year and a half considering the sexual behavior of dogs and rats. But he began to think he was in graduate school mainly because it provided a deferment from the draft, and so he had one of those predictable crises: *Why am I here?* He liked California. He was excited to be in graduate school. But the research—dogs and rats fucking—was starting to get old. For a change of pace, he took a course in the anthropology department at a time when one of the professors, Phyllis Jay, said that Jane Goodall wanted someone to study baboons. Jay had earlier done a field study of monkeys in India, and she knew Jane slightly as a colleague. No one in the anthro department was prepared to go to Africa and study baboons, though, so Tim from the psych department volunteered.

He had flown out to meet Jane and Hugo at their camp in the Ngorongoro Crater near the end of September 1967 before proceeding on to Gombe. By then he had been thoroughly Californianized in the Berkeley style of the late sixties, which meant, among other things, that he arrived in the crater with a Smith Brothers beard and Jesus Christ hair. Jane, aware of some expressive new fashions in dress and grooming, was nevertheless shocked by the sight. Writing to her family in England that September, she announced, "Tim is a real, live HIPPIE!!!!! He is not wearing his beads, but he has them with him. He is nice none the less—but his hair is GHASTLY. And he has a beard. And he has just got married."

The married part was in his favor, and by the time Geza appeared on the scene about six months later, Tim's wife, Bonnie, had arrived, while

Baboon meets mirror.

Tim's field study had, in another sense, nearly arrived. By then—late March and heading into April 1968—his study was starting to rival those of his only established predecessors: two or three earlier baboon watchers who had either contemplated the creatures while sitting comfortably in a safari car or, if approaching on foot, had seldom moved closer than twenty-five yards from their subjects. Tim was already getting closer than that, and within another few months he would be able to walk right in the middle of the troop and be mostly ignored by them.

But Tim had not yet reached that level of trust with the baboons. He had just about figured out how to tell them apart, and even that was not easy. There were more than fifty individuals in Beach Troop. The easiest to identify were the adult males, because there were so few of them and because their personalities showed through quickly. He was identifying them on the basis of style and character, of personality, and he gave them names meant to reflect those personalities. Names like Harry and Myrna, Crease and Chip, Moses and Quasimodo, Will and Thug and Grinner. In fact, he and Geza had long discussions—arguments, really—during dinnertime about whether that was a valid approach. Do animals even have personalities?

At first, Tim found Geza mildly intimidating. The intensity. The divergent eye. The hatchet face. Tim's arguments with Geza had mostly to do with the us-and-them problem: that whole stupid issue of anthropomorphism, the danger of overprojecting the emotion-laced minds of the observers (humans) onto an alien reality of the observed (baboons). Was Tim being scientifically objective? Was he imagining human qualities in his baboons that weren't really there? Geza was ready to argue about those things, but Tim responded quickly and emphatically: The baboons had plenty of humanlike qualities. Baboons are primates. People are primates. Tim wasn't about to draw a line and say they differed entirely from people in any particular behavior.

7.

Gombe's isolation could give a person the sense of living in a world apart. The regular world, the civilized world and its steadily expanding catastrophe, could seem like a far-off dream, its reality marked as the faint and distorted signal sent by BBC World Service and Voice of America or inadequately summarized in the flimsy, tattered scrolls of *Time* and *Newsweek* arriving weeks out of date with the rest of the mail on the Kigoma supply run.

As a result of such imperfect communications, Geza heard about the April 4 assassination of the American civil rights leader Dr. Martin Luther King two weeks after it happened, when he opened a letter from Ruth to read about the tragic event and its depressing sequel, the riots that raged in Washington and elsewhere in big cities across the country. There was a personal aspect to the news. Ruth, living as she did on the fringe of one of Washington's bitterest neighborhoods, was hurt on her way home from work. Her letter was no more specific than that, which was typical of Ruth. Big private life, small public one. She was reluctant to describe the personal details, even for Geza. He had warned her about the dangers of where she lived, in that seventh-floor apartment of the Dorchester House on Sixteenth Street. But along with being private and self-contained, Ruth was also stubborn. She didn't listen. She needed to save money, and the apartment was a decent place that rented for a good price because it was next to a bad neighborhood. He had told her to find another place. "I'm staying," she said.

Even then, from the distance of Africa and the remoteness of Gombe, he could picture Ruth and the furniture and layout of the apartment. It was where he had stayed for days on end, after all, and where they said good-bye when he left for Africa. She didn't want to go to the airport, so they parted at her door. They would probably never see each other again, or so it seemed. Ruth expressed no regret, no desire to hold on. She encouraged Geza in his going off to Africa, and she made it clear that she was not expecting a future relationship.

. . .

Ruth and Geza had both graduated from George Washington University by then, but while he went off to study anthropology in graduate school at Penn State, she stayed in the city hoping to save money for graduate school at some point in the future. She had been a geology major, and she wanted to do graduate work in the same subject. Now, though, she worked as a typist for a Russian-English translation service run by the father of Geza's friend George Rabchevsky. She got the job because she was a skilled typist, but it didn't pay well, and she had no promising prospects for the future. In sum, Ruth was living in limbo: perpetually short of money and believing that she might have to go back and live with her parents in Virginia.

But then, within a month after Geza's flight to Africa, he was able to write her and hint at the possibility of work at Gombe. He was still staying with the van Lawicks at their camp in Ngorongoro Crater at the time. Jane and Hugo had been talking about their hopeful plans to develop Gombe as a major scientific research station and their need for more people. Jane was terribly overworked, trying to do several projects at once, and she wanted someone who could type. Ruth was, Geza claimed, the fastest typist in Washington. She was fast. And while she never did become a typist at Gombe, that was the premise on which the invitation was first made, although Jane also wondered if Ruth could handle babies. Would she babysit?

At first, Geza wrote to Ruth tentatively about the idea: "Do you think, if there's a possibility of coming out here, you could come soon?" But before he got an answer to that one, Jane had confirmed the fact that she wanted a typist and someone to help take care of Grub. Geza wrote another letter saying, in essence, "If you're willing to come, get ready to come right now because I don't know how long this offer's going to last."

In speaking with Jane and Hugo, Geza had been honest about Ruth and the nature of his relationship with her. They were friends and lovers, and he missed her, but they were not ready to think about marriage. Jane accepted Geza's description of the relationship, but, perhaps to satisfy other people's sense of propriety, including that of her rather Victorian grandmother back in England, she persisted in describing Ruth as Geza's fiancée.

Meanwhile, Geza was trying to reassure his mother and father, also George Rabchevsky, on the propriety of bringing Ruth out to Gombe. His mother was her usual impossible self: concerned above all with the undeniable fact that Ruth was not an aristocrat. His mother wrote to Geza in Hungarian, but "bloodline" and "breeding" were how the critical words in Hungarian translated, as if she thought they were all horses. His father and George were more reasonable if still irritatingly conventional, concerned as they both were that this new development might mean a premature parachuting into matrimonial territory. George's reaction upset Geza the most, since George was a friend and ought to have understood. For the "sake of our friendship," Geza wrote to him on April 18, trying to explain the situation and Jane and Hugo's attitude. He had never "tricked" them into agreeing to take Ruth. They were looking for more volunteers to help with the chimpanzee work, and when he told them about his relationship with Ruth, they came up with the idea of bringing her there. It did not mean that he and Ruth were planning to get married, and there would be, after all, other people around, so it was not a matter of the two of them being alone together in the jungle!

. . .

Money had always been a concern for Ruth. When she started as an undergraduate at George Washington, her biggest problem was paying the tuition. She had taken various odd jobs, waitressing, and so on, in an attempt to do that. Once she became a geology major, Geza's father arranged for her to take a half-time job in the department working for a mineralogist. She learned to peer beyond the limits of light by operating an electron microscope, and so she became a special assistant to that person. Although it paid poorly, the job qualified her for a tuition waiver at the university. It also brought her into the geology department for four hours every weekday.

Geza was by then coming to the department regularly, not so much because of his father, who headed it, but more because of his friends, who were graduate students there. Geza's father took an interest in his students well beyond what a normal teacher would do. In a sense, he adopted people. He did things that helped them out in their personal lives as well as in their academic careers. These adopted students were usually people who needed help—some were impoverished foreign students from Russia, Germany, the Middle East, and so on—and several developed informal relationships with him. They hung out around the department, came to the house, had meals there, and so on.

His father had established a field station in West Virginia, where students came out for weekends to experience practical geological work. Geza's parents had divorced years earlier, but then his father remarried, and his stepmother loved to cook. She would come along and cook for everyone for two or three days, and they'd all be sleeping in the same room. It was that kind of environment, and it produced a group of students with similar backgrounds and interests and a lot of camaraderie. Geza was younger than the others, but he had always gravitated to people older than he was. So he went along on those trips and made geologist friends; and whenever he was on campus, he was likely to drop in at the geology department. That's where he met Ruth.

She was quiet. Reticent. Not sociable on the surface. She had a way of looking at people, including Geza, with an absolutely direct, straight-on look. Not a stare, but more the look a chimp gives. No hiding. No eye-shifting. No tendency to act like she wasn't looking. She would look with an open face that showed nothing more than curiosity. She also did not laugh or smile, except on rare occasions, when there would be the faintest expression of a smile on her mouth or in her eyes. But it would not be vocalized or expressive or outgoing. It was a private smile, as if the smile were meant for herself and no one else. You had to draw her out by asking something or directly approaching her. It was, in fact, a couple of years before Geza seriously interacted with her. And when he first really noticed her—small and trim with lustrous auburn hair—and began to see that she was attractive and interesting, and considered that he might try speaking to her, he was afraid to. He had gone to boarding schools all his life, which made him one of those socially delayed adolescent males. He didn't know how to interact with females. Girls—or women—made him nervous. So it was a great surprise when, one day, he asked her to the movies and she agreed to go.

They went to see a Fellini movie at one of the art cinemas on Pennsylvania Avenue near Twentieth Street. It was a full-length black-and-white film called *La Strada,* starring Giulietta Masina, Fellini's wife, along with Anthony Quinn, and it focused on the story of a small, itinerant circus peopled by crazy characters. At least Geza thought the characters were crazy. In fact, he couldn't figure out which segments were meant to be real and which surreal. It seemed like an exercise in madness. He walked out of the theater shaking his head, convinced that they had just wasted time and money on nonsense masquerading as art. He said to Ruth something to the effect that a person had to be out of his mind to make a film like that. But Ruth had understood it entirely, and she spent the next fifteen minutes explaining the film to him. All the complicated psychological bizarreness in the film made sense to her, and her ability to understand it so easily and explain it so directly caught his interest. She, meanwhile, became frustrated with his incomprehension, which she dismissed at last with a wry smile and a wave of the hand. He would always remember the smile and wave, and he saw that he was dealing with someone unusual.

. . .

After that first date, they spent a year in a slowly developing relationship. You didn't push Ruth, and the relationship was not initiated by sex followed by affection. It developed in the opposite way, as a quiet sort of mutual interest gradually followed by attraction and then intimacy.

They began doing social things with others from the geology department, most often George Rabchevsky and his girlfriend, Olga. George's family were White Russians who had fled the Soviet Communists at the end of World War II, just as Geza and his family had, which was a formative experience the two men shared. Sometimes the four of them would take Geza's dad's 1962 Cadillac, with Geza at the wheel of that five-ton, two-toned, white-sidewalled behemoth. It was big enough to haul eight students and a trunkful of geologically interesting rocks back from West Virginia, and when Geza drove it he had trouble seeing over the steering wheel.

They might drive along the Potomac to Fletcher's Boat House, rent a boat or a couple of canoes, and spend an afternoon on the river. By the time they got back, they were exhausted and sunburnt. Other times, they would head up to the Glen Echo Amusement Park, which was antique even then, with its art deco entrance and a roller coaster that, unlike any other in the country, carried you through the middle of a forest so you couldn't see where you were going. The place also had rides like the Spinning Tub (you were supposed to sit down, but people kept trying to get up), the Whip (violently spinning carts), and bumper cars (hard-to-drive miniature electric cars that crashed into each other while the sparks sizzled above). Those were mainly rides for adolescent boys, and for Geza and Ruth, George and Olga, going to Glen Echo mostly meant walking around, participating in assorted games of chance, and eating terrible things like cotton candy. But Ruth also liked the old-fashioned carousel. That was her favorite. And there was a huge swimming pool with a wave machine, possibly the first such device in the country. Since the pool was usually not crowded, they might go there to cool off. Or they might try dancing at the dance hall. The hall had mostly big-band dancing from another era, though, which Geza and Ruth were not interested in.

Music. Geza first heard a Beatles song while sitting in a bar on Pennsylvania Avenue in downtown Washington. It was strange and nervous with falsetto riffs yet light and positive: "I Want to Hold Your Hand." But his musical tastes had already crystallized by then, and he never became much of a rock 'n' roll fan. By just a tiny sliver of time, just a year or less of life experience, he was pre–rock 'n' roll. Ruth had gone through a phase of being crazy about Elvis, but the music he and Ruth listened to together was mainly the old-style pop, a sweet and ultimately reassuring strain of harmonic fashion rising out of the postwar fifties and early sixties: the Kingston Trio and the Brothers Four, the Coasters, Everly Brothers, Harmonicats, Herb Alpert and his Tijuana Brass, Duane Eddie and his twangy guitar, Bo Diddley, Sandy Nelson, Julie London. That was the music they liked and identified with, the melodic memes implanted in their brains during the first fertile flush of sexual maturity.

They smoked cigarettes. Menthol brands for Ruth. Viceroys for Geza. But they never touched marijuana. Never thought about it. Drugs! They drank an average of about one bottle of wine a year. They were innocents learning to swim in the river of life.

• • •

Sex came slowly, intimacy at its own unprodded pace. The first time Ruth invited Geza into her bedroom happened more than a year after their first date. By then, she had moved from an earlier place in Georgetown to an apartment on Vermont Street near Thomas Circle. The apartment building, known as Crescent Towers, was just off Fourteenth Street. At the time, Fourteenth Street was the sleaze center of Washington: liquor stores, porno shops, hookers on the corners. She had left the Georgetown apartment because she could no longer afford it, and she was now renting something cheaper. It was her next-to-last apartment in Washington, the one before she moved into that final place on Sixteenth Street.

Ruth cooked a dinner for Geza at the apartment, and she was describing a new job she had taken, some kind of waitressing job. She told him

Ruth in her apartment off Fourteenth Street in Washington, DC.

about it but not with much detail, and so at first he misunderstood. It turned out that she was waitressing drinks in an imitation Playboy Club, and she had to wear a sexy outfit as part of the job. Geza said he thought he wanted to see where she worked. She said she didn't think that was a good idea. He said, "Is it that bad?" She said, "It's not always pleasant, but I make a lot of money."

She took out the costume, which was in her bedroom closet, undressed in front of him, and put it on to model. It wasn't topless, just skimpy, like what Playboy bunnies were wearing in actual Playboy Clubs. Mostly front display. Some cleavage. Typical of the sixties, and Ruth filled it out perfectly. She had the body for it. There was no skirt. The bottom part was like a high-cut swimming suit, and mesh tights covered the wearer's legs.

Ruth took risks that made Geza uncomfortable, and one of them was serving drinks in that imitation Playboy Club located in one of the worst parts of town. She did not act like a sexy doll. She was not flirtatious. But she could transform herself in a few minutes, in ways that were a total mystery to him, from an average schoolgirl to an attractive, full-blown woman. A twist of the hair, putting it a certain way. It was a trick she had.

So they embarked on the experience of physical intimacy, and their relationship after that became deeper and more stable. But it was still not easy to get to know Ruth. She was very much her own person, and for that reason some things she did really shocked Geza. One time, for example, she took him out to see a pornographic film. It was her idea, not his. She suggested it without any discussion or warning. It was just: *Let's do this.* They walked from her apartment past Dupont Circle to a theater at the corner of Connecticut and R. The theater showed a variety of films, art and underground films, and this particular movie had been produced by the Playboy company as a mass-market film. In other words, it was pornographic but with a small pretense of respectability and some plot. It was based on a biblical theme.

Geza never learned to be casual or overt about sex, and it was with a mixture of curiosity and genuine discomfort that he walked into the dim theater. Ruth didn't say a thing during the entire film. He would sneak sidelong glances at her at various moments, but she seemed completely involved in watching everything. The film was definitely hardcore, and he observed people do things on the screen that he had never before even imagined were done. Then they came out of the movie, and he was afraid to say anything. She didn't say anything either, and they went back to the apartment, where she wanted to try some of the things they had seen in the movie.

8.

Some people said the baboons were a problem, but Tim Ransom never considered them such. Why should he? When you thought about things from a baboon's perspective, you saw that they were living out their lives with passion and calculation. They made choices. They laid down plans. The baboons were not even very aggressive. Because a few earlier researchers had overemphasized the importance of aggression in baboon society, Tim began his research with that idea in mind, wondering what all this aggression business was about. He would think to himself: *I better pay attention to who's going to beat me up for being in the wrong place at the wrong time.*

It took him a while to realize that their aggression was situation-specific. There were different kinds of aggression, and it was all very stylized. True, cuts and fractures occurred commonly enough, yet these animals were all so quick and capable that only rarely did dire things happen. Usually there would be a slice from someone's sharp canine or maybe a broken finger or two. Tim began to realize that male-against-male aggression was almost never done with the intent of engaging in an all-out fight to the death. Among the males, fights usually had to do with getting access to females and to friends. Fights among females were usually about access to babies or food.

Unlike the chimps, who never stayed in one coherent group but were always forming temporary subgroups, then splitting away to form other temporary subgroups, the baboons always stayed close together in their home troop. There were exceptions, the most obvious being when an adolescent or young adult male transferred from his home troop and joined another. This would be a major event and—as with a human late-adolescent leaving his home in, say, Berkeley, California, to do primate research in Africa and finding himself in unfamiliar territory surrounded by strangers—very stressful for the individual. At the same time, the arrival of a new young male would send ripples throughout the social network of the troop, with shifting relationships and alliances causing a readjustment in the matrix of power and status.

Aside from those periodic transfers, the whole troop, all fifty or more individuals, kept continuously in touch with one another. Baboons in a troop could always hear or see one another. They moved as a group, went through rambling perambulations that began in the morning at the twenty trees they slept in and ended in the evening at the same twenty trees. This daily stroll took Beach Troop along the beach; and it took the second study group, Camp Troop, through the banana-provisioning meadow at upstairs camp. In both cases, the baboons proceeded in an approximate circle, with the greatest unpredictability in their movements being whether they would do it clockwise or counterclockwise. This was Baboon Economics 101: the daily search for food and water, sex and status, family and community, and a safe tree to sleep in at night.

To know about the baboons and to live among them, you had to learn their language and thinking. Eventually, Tim came to imagine that the baboons regarded him as a strange version of a subadult. *Subadult* was what he called the older juveniles, those on the cusp of being physically

Tim on the beach with Beach Troop.

mature but still considered youngsters by the others. They were gangly adolescents, tripping over themselves and making a lot of social mistakes. As someone slowly learning the baboon language—gestures, postures, vocalizations, and so on—Tim, too, was making lots of social mistakes, and so he began to see that the adults in the troop were treating him as just another bumbling subadult. They tolerated him because nobody took subadults seriously. He wasn't a threat to the adult males, competing for access to the sexually receptive females. He didn't threaten the infants and juveniles, so he never got into serious trouble with the adult females. And that's how it went during his first year of hanging around with Beach Troop, slowly working his way in, bit by bit picking up the language and appreciating the immense complexity of those animals.

It was wonderful. So was being there on the beach all day with the Beach Troop. Tim's father frequently got transferred as a result of his job when Tim was young, and thus Tim went to thirteen different schools before high school. The idea of staying in one place, getting to know what was under every leaf and rock, and learning to recognize each one of fifty nonhuman primates by name and personality, was an intense and reassuring thing. The relationships were fascinating, and the drama of the baboons became the most absorbing thing in the world. It was like watching the best possible TV series, although what Tim liked most about baboons was that they never lied. He was going through some personal issues having to do with trust, and that led him to think of baboons as his favorite people. You could always tell what they were up to. To be sure, knowing what they were up to was a consequence of recognizing their gestures and vocalizations and understanding their personalities. Knowing, for example, as all the baboons in the troop did, that when adult male David threatened he never intended to follow through.

. . .

Bonnie, Tim's recently arrived wife and colleague, hiked upstairs to the main camp every morning to study the other baboons, the ones in Camp Troop, and she would have had a different kind of experience, one undoubtedly colored by the regular conflicts between baboons and chimps at the provisioning site. True, the banana operation was originally put in place to entice the chimps, and if baboons could read, someone might have put up a sign that said No Baboons Allowed. Since they couldn't and no one had, who was to say the baboons were at fault? They had discovered a reliable food source and were naturally trying to exploit it.

But it was stressful for them to be up there in camp next to the chimps. Of course, baboon youngsters often played with the chimp youngsters, and those mixed playgroups, cute little stump-tailed apes and cute little whip-tailed monkeys leaping and spinning about, chasing one another round and round, pretending to fight with each other, were endlessly delightful to watch. But bigger chimps would sometimes mug smaller baboons and take their bananas, while the bigger baboons would sometimes gang up on smaller chimps and grab their bananas.

It was getting to be a free-for-all, with the worst part of it being a consequence of the dark truth that chimpanzees sometimes eat baboons. Baboons are a kind of monkey, after all, and the chimps were skilled hunters and ferocious predators of monkeys. The chimps loved meat, and they sometimes took great risks to get some. So it was odd that life could sometimes be so peaceful, with juvenile baboons and juvenile chimps playing together like two different kinds of primates frolicking in Eden, whereas at other times the younger baboons might suddenly become just one more entrée on the menu.

. . .

Once, at 8:15 in the morning in the middle of March 1968, it happened like this. Arwen Baboon was sitting on top of a closed banana box and holding in her loving arms her little infant, Amber. Nearby was a big tree, and sitting in that big tree, a few feet above mother and baby baboon, were three adult chimpanzees—Mike, Charlie, and Hugh— peacefully grooming one another.

Arwen the mother baboon was facing the tree as she sat, holding her baby while looking up at the three big chimps grooming. Nearby on the ground sat another chimpanzee, Leakey, who was steadily eating a clutch of bananas at one of the opened banana boxes. Leakey was still focused on the bananas when Figan Chimpanzee suddenly appeared out of the forest and approached with a big toothy grin that was probably an expression of fear. For unclear reasons, Leakey dropped the bananas and jumped on Figan, and the two of them began rolling down the slope in a wrestler's clutch, with Figan screaming wildly.

Mike, still up in the tree and grooming with Charlie and Hugh, may have been upset by this event because he was the alpha, the big boss, and two of his subordinates—Leakey and Figan—had the nerve to challenge his authority by fighting right in front of him. He turned and began swinging down and out of the tree. He was about to display, to show off his mighty might by doing the weight lifter's ballet. When he displayed, Mike liked to throw or flail things as a way of emphasizing the drama. On his way out of the tree, therefore, and while still hanging on to a branch, Mike reached down and grabbed with his right hand the nearest convenient thing, which was Amber, Arwen's baby. He landed on the ground and went into a full display: standing upright, his hair raised and spiking out, then running while screaming and flailing at the ground with the object in his hand.

Baboons are communally protective of their babies, and so at the sight of Mike running about and hitting the ground with Amber, several of the big male baboons, their bright teeth bared, converged on the big, hairy ape from all sides: barking, screaming, grunting. Others in the troop added to the chaos with screams and squeaks and squeals. In response to the aggression of the baboons, some of the male chimps began to move in, barking and screaming, while the female chimps, some holding infants and grinning in fear, added to the commotion with their own screams and squeaks and whimpers.

Possibly in order to keep both hands free while defending himself from the sharp-toothed baboons, Mike put Amber into his mouth, gripping the limp infant by her back. Surrounded by a scrum of excited baboons and chimps, Mike then paused, placed Amber in his left hand, and began running downslope toward a thick patch of tall grass at the bottom of the meadow while reaching up to pull off a big baboon who was clinging to his back and biting his shoulder. In the process of ripping the baboon off his back, however, Mike whipped his left arm back and accidentally smashed his left hand into a tree trunk, whereupon Amber went flying and disappeared in the tall grass.

With the baby suddenly gone, the commotion quickly subsided. Mother Arwen, looking dazed, began wandering about, searching, it seemed, for her lost baby. Other baboons also began moving into the thick grass, apparently joining in the search. Some of the chimps did as well. Soon, in fact, individuals of both species were acting as if they had forgotten the entire uproar, and they milled around side by side in the tall grass, all of them seemingly intent on looking for little lost Amber.

There were still brief flashes of interspecies hostility. At one point, Cyrano Baboon grabbed the hair on Rix Chimpanzee's back, yanked out a big clump, and barked aggressively. Rix turned around, stood upright with bristling hair, and slapped Cyrano on the head. Cyrano ground his teeth and slowly blinked, displaying his bright white eyelids. He bounced up and down on stiffened front legs. Then big Hugh Chimpanzee became more generally upset and began stomping the ground and flailing with his arms. He turned and, perhaps not seeing any better target, charged directly at the new human observer on the veranda— Geza—who jumped into the air at the last minute, so that Hugh missed contact and continued racing through the meadow.

An hour later, one of the human observers spotted Humphrey Chimpanzee, who had not taken part in any of the earlier melee, sitting peacefully by himself in a tree on the other side of the valley. Humphrey was chewing slowly on a sticky rag of meat that looked very much like the bloody remnants of little Amber.

9.

In the final week of May, Ruth, accompanied by a new short-term volunteer from the Friends World Institute in Nairobi, climbed into a small plane at Wilson Airfield in Nairobi. A few hours later, they were met at the Kigoma airstrip by two people from Gombe. Patrick McGinnis was one. The other was probably Mpofu, a young Tanzanian who had been hired as the Gombe boatman and general carpenter. After finishing the weekly shopping in town, all four walked down to the dock and loaded themselves and their luggage and supplies into the Boston Whaler. There was a bobbing passage through flashing sun and crystal water until, after a time, the boat stopped at a tree-shadowed beach where Geza stood. He must have looked leaner and darker than Ruth remembered. She may have felt weaker and lighter and become temporarily less stable in response to the repeated impress of shifting balance and the novelty of everything, including Geza standing at the edge of a vast African lake, Geza stepping into the water to catch the boat and draw it to solid shore, Geza catching her hand.

Mpofu attended to the boat, while a couple of other African workers who had run down to the shore lifted out the food and some of the other items bought in Kigoma. Patrick and Geza, carrying much of the luggage, turned and led the way upstairs, while Ruth and the new volunteer followed. They entered a flickering forest and were beckoned up a winding path by waving hands of light, arriving finally at the clearing and the main camp, which was glazed yellow by the late afternoon sun. Ruth noted two rectangular aluminum buildings resting on concrete slabs and covered by dark thatch.

No chimps or baboons were in camp, and so, without trans-species distractions, they all dropped what they had been carrying inside Pan Palace, and Geza began heating water for coffee and tea. Carole, who had been sitting at the table in front of a typewriter, introduced herself to Ruth and the new volunteer, promising to help the latter find her way to an empty hut that would be hers. Alice Sorem showed up, so now there were six of them, and they all sat down for a brief introductory chat before at last heading down to join Tim and Bonnie for dinner on the ridge.

After dinner, Geza helped Ruth settle into the extra room in Pan Palace, where he had been living, and he then showed her his new hut, which was a bright aluminum rondavel recently erected along a narrow, twisty path running into the woods away from the upstairs camp clearing.

. . .

Ruth was secretly terrified on her first morning there, which was a problem Geza would not recognize until, decades later, he read some of her letters home. She lay in her bed inside Pan Palace that morning for two hours, inert, huddled under a blanket, drifting inside the white cloud of mosquito netting, and listening helplessly to the histrionic racket of chimps outside. She thought about ways to catch a plane back to the States.

By the time she wrote her first letter home to her parents, on June 4, she had been there for about a week and, during that time, had overcome the worst of her earlier fears. "How remote the rest of the world seems!" the letter began, and then: "Here we are in the world of bananas and chimpanzees and nothing else seems real." She was "slowly getting into the swing of things," she added, and she had already spent time in the provisioning area when things were "relatively calm," learning to observe behavior and tape-record notes. Still, she wasn't sure that she would ever learn to tolerate a "banana morning" at those times "when there are 30 chimpanzees and 50 baboons fighting and carrying on like crazy!"

The chimps themselves created pandemonium enough, Ruth admitted to her parents, but stirring baboons into the mix was asking for trouble. The baboons squabbled with each other. They fought with the chimps. Altogether they made the situation tense. Those doglike monkeys with enormous snouts, bright white eyelids, and pug-ugly bottoms were simply "repulsive," she wrote; but once they had wandered off for the day the chimps could settle down and enjoy themselves, and the whole operation was actually a pleasure to experience. She so wished her parents could see the chimpanzees. They loved bananas, even though most of them were surprisingly fussy: discarding fruits that were too ripe or too green, concentrating on ones that were just right. Ruth had tried the bananas herself. They came in an astonishing variety, and the good ones were tasty indeed, so who could blame the apes for being obsessed by them?

The regular food that people ate ran the gamut from excellent to "extremely gross." For breakfast and lunch, they had eggs, jam, honey, cheese, sometimes bacon, and bread—with the bread being the best surprise, since it was baked fresh every other day by Sadiki Rukumata, the cook. They prepared their own breakfasts. The eggs tasted like sardines, since the chickens were fed the silvery, sardinelike fish that the lake fishermen scooped up; with the distant hope of masking that fishy taste, everyone scrambled their eggs and added a few drops of vanilla extract. Oddly enough, fresh fruits and vegetables were in short supply, although tangerines were just then in season and therefore plentiful. But the daytime food was the excellent and good part, while the more questionable and sometimes extremely gross part arrived in the evening. True, fresh fish was served on Thursdays, the usual day for a supply run into town. But on the remaining six nights of the week, dinners were structured around meat dug out of cans. Sometimes, it appeared as meat stew. Not too bad. Other times it was transmogrified into a meat pie. Remotely edible. But then the canned meat would finally be reconstituted in a spiced curry smothered with fried bananas, which was simply disgusting.

Everyone craved dessert, and since Ruth had recently been elected the dessert cook, she had just that day baked a cherry pie. Unhappily, Sadiki's charcoal oven broiled rather than baked it, and the crust had been burned black in fifteen minutes. Ruth expected to develop her culinary skills over time, but for the moment she hoped her mother would send recipes for desserts that did not require baking.

. . .

Back in the United States, on the same day Ruth wrote that first letter home—June 4, 1968—Robert Kennedy, younger brother of the assassinated President John Kennedy, won the California vote in the Democratic primary for the forthcoming presidential election. Late that evening, Senator Kennedy effusively thanked his staff and supporters during a ballroom rally at the Ambassador Hotel in Los Angeles. Some time after midnight, as the senator was being escorted out of the ballroom, passing through the kitchen and heading for a rear exit, an aggrieved if otherwise unremarkable young man pressed forward and at close range shot the candidate three times. He died a day later.

In her next letter home, written on June 10, Ruth referred to the second Kennedy assassination briefly, commenting that "I can't imagine what the poor U.S. is coming to." She then described her own sense of placid remove from the madness that had lately seemed to overtake her home nation. Living so close to the natural world, as they did at Gombe, made Ruth think that the life she used to live in the States was artificial, and she did not miss it. She missed only her dear mother and father, and it would be wonderful if she could fly home for a day or two, just long enough to see them and describe in detail "the wonders of this life." Those wonders included the chimpanzees, of course, and Ruth now was becoming "so fond of them!" The chimps seemed to have such a thrilling existence, and indeed, she envied them. "How beautiful it would be to build a nest at night high in a palm tree and lie in it under the African sky with the lake below and only the sounds of the wild African night!"

In fact, Ruth continued, she loved living in the middle of an African forest surrounded by so many wild animals, and she was even delighted by the camp's bathroom facilities, the most fundamental aspect of which was the *choo*. The choo was a horseshoe-shaped seat elevated above a hole in the ground and isolated on three sides by airy walls of stick and thatch. The fourth side was open, and there was no roof, so a person at night could contemplate a drifting moon or the bright sparkling dome of stars overhead and, during the day, enjoy a superb view. The bathing area of the facilities consisted of a cool, swelling pool near the bottom of the small stream that ran past the camp. The pool was clear and the place private, so taking a bath there was a "delicious" experience that made one want never to use an ordinary bathtub again.

. . .

Ruth wrote home again on June 22, describing in detail the nature of her work as an observer and adding the comment that "for reasons that I will elaborate later, our observation procedure has been altered somewhat in the past week or so."

The procedure for researchers doing observations relied on using bananas as bait to get the chimps coming into camp regularly, and it seemed like an ideal way to keep track of the apes. It had started simply. When Jane first went into the forest in 1960, she managed to sight chimps only occasionally and from a great distance. She was getting closer within three or four months, and within a year, a few of the more curious chimps would occasionally wander into her camp. To encourage those cameo appearances, she once or twice tried leaving out a few bananas. That's how it started. After a time, she left out more bananas, and the curious chimps began coming into camp more often. But then they began taking things other than bananas. Old shoes. Patches of canvas. Cardboard. And when Jane tried making the bananas more challenging to find, hiding them as if they were Easter eggs—in trees, inside various containers—the apes responded directly. Shoes were torn open and eaten. Coffee thermoses and flashlights were smashed to see what was inside.

After Jane and Hugo were married, in the spring of 1964, the two of them worked to regularize the banana provisioning. They brought in steel boxes with lids that could be opened or shut remotely, using levers and cables, with pins to fix the levers and cables in position. One of the chimps began breaking the cables. Jane and Hugo put them inside steel pipes, which were then buried underground. After a while, three chimps discovered how to remove the pins that kept the boxes shut. Jane and Hugo replaced the pins with screws, but the same clever trio figured out how to unscrew the screws. Jane and Hugo ordered better boxes, forty of them, which were made from steel, embedded in concrete, and locked with electrically operated latches activated remotely by push buttons.

Those were the boxes being used when Geza and Ruth arrived. By then the baboons, too, had discovered the bananas. At first, they had been afraid of people and kept their distance. It used to be that a person could threaten the dog-faced monkeys simply by looking at them directly with a meaningful gaze or an assertive stare. When that kind of threat lost its power, people learned to raise their arms into the air and lunge. When that wore out, it became necessary to throw stones. At the moment, only one person in camp was still able to control the baboons: Geza. Ruth was much smaller than Geza, and for her, being out on the slopes and surrounded by the baboons was no fun at all. In fact, she now had the distinction of being the only person in camp to have been attacked by one. That happened at the start of an afternoon when the banana feeding had ended. All the chimps had wandered away, so it seemed like a peaceful moment. Ruth was standing outside when a big male baboon swaggered past, perhaps only eight feet away. She thought nothing of it. But then he spun around and leapt on top of her. She started to fall, then regained her balance while throwing her arms up and lunging forward. At that point, the growly beast moved a few feet away, but he was still there, and it had been an exceedingly unpleasant experience.

Even before Ruth showed up, Carole and Geza had both started to recognize the seriousness of the baboon problem. Geza could throw a stone. Even with his one bad eye he had a better aim than anyone else. Carole once threw a stone at a baboon, who just watched the projectile spin his way and then, as casual as anything, twisted himself sideways at the last second. Carole didn't have the muscle to throw stones fast and hard like Geza did, and she couldn't use the camp's slingshot now that she was an observer who tape-recorded notes, since she needed one hand free to hold the microphone.

Then came the attack on Ruth, so finally Geza approached Carole and said, "I think this banana feeding thing has gotten out of hand, and I want to shut it down." He began to list, one after the other, several logical reasons why they should do it. Carole had been seeing the same problem he had, but she hadn't had the sense that she could change it. The instant Geza proposed it, however, she said, "I'm with you. I agree. We should shut the feeding system down." Patrick didn't seem to like the idea. But Geza, Ruth, and Carole went downstairs to talk to Tim and Bonnie, who had also become critical of the banana provisioning, albeit from a baboon-watcher's perspective. And that was that: a rebellion in the ranks. They shut the system down.

Patrick radioed Jane and Hugo to report on the week's events, and they, having heard little or nothing about the developing problems with baboons, were surprised and upset. Geza then got on the radio and strongly argued his case. They would use up the last of the banana stockpile, he said, and then, instead of continuing the provisioning, they would try following the chimps into the forest. Jane and Hugo were about to come out for the summer, and so they all agreed to wait until then before attempting any further changes in the routine.

III

The Golden Summer

(June to September 1968)

1.

The Land Rover was precariously overloaded and drawing a trailer. Then a spring on the trailer broke. Then the transmission gave out, or seemed about to, until after a hundred miles of gentle nursing, with Hugo shifting as seldom and carefully as possible, it seemed to fix itself.

Hugo was staying about five minutes behind the gray VW bus, which was far enough back for the dust to settle. Nic Pickford was at the wheel of the VW, his wife, Margaret, sitting next to him, and—seeing the tractor ahead moving much too fast, bouncing madly and weaving erratically—he knew there would be trouble. Standing next to the driver of the tractor and hanging on for dear life was an African woman, very pregnant; a young boy was seated on the driver's lap. Then the driver abruptly pulled over to the side of the road to have a quick chat with someone, and Nic drove past.

Hugo, Jane, Grub, and yet another short-term volunteer from that Quaker school in Nairobi were all tightly packed into the overloaded Land Rover, and by the time they arrived at the spot, the driver of the tractor had started up again and was once more going dangerously fast. The woman, having seen the Land Rover move up behind them, shouted something into the driver's ear, and he, without slowing down, pulled over and lost control. The tractor spun and bucked, and the driver and his two passengers were thrown off. One of the great rear wheels of the thing ran right over the man's chest, then over the woman's head.

The driver quickly recovered, dashing after his machine, climbing up and turning off the engine. The pregnant woman stayed where she was, lying on the ground and jerking convulsively. Hugo stopped the Land Rover, grabbed a blanket, and jumped out running; but before he could reach her, the woman was up and staggering aimlessly, blood oozing from her mouth. He tried to persuade her to lie down once more. The boy was also on his feet, having sustained a couple of scratches and the loss of a tooth. The man was walking in the road now, unsteady, spitting blood.

A car pulled up, its driver a German missionary who, after some urging, agreed to take the man to the local hospital. He knew where it was. Jane and Hugo helped the woman and boy into the Land Rover and then caught up with Nic. The VW had more room and better springs, so the woman and boy were helped into the bus, and they followed the German missionary on to the hospital. Or, rather, the supposed hospital. It turned out to be nothing more than a tiny clinic, but they were hours away from anything better. At the clinic, Nic quietly told Jane that he thought the woman wouldn't survive and that there was nothing more they could do. It was a nightmare, and it left Jane feeling sad and uneasy, more fatigued and out of sorts than she had been to begin with.

. . .

They reached Kigoma in midafternoon on Sunday, June 23, dead tired. They parked the two vehicles, hired a water taxi, loaded in all their things. Grub loved the boat ride. He was wide awake and excited to be riding on the lake, although his exuberance was temporarily dashed when the lake swallowed his favorite toy—but soon the tears dried up, the lost toy was forgotten, and the sunny disposition returned.

Nevertheless, arriving at Gombe was a distinctly unpleasant experience, as Jane had expected it would be. All the chimp news was "grim, grim, grim"—so she wrote in a July 1 letter to the family (mother, grandmother, two aunts, one uncle) back in England. A flu epidemic had killed a number of the chimps, including Jane's favorite, David Greybeard, and dear Faben had not been seen for a long time. Then, of course, Jane also had to deal with the baboon and banana problems, as well as those unhappy researchers. As she wrote, "Baboons attacking humans. Baboon had a tooth broken by Geza throwing a stone. Ruth attacked. Carole wanting to leave. Gloom. Hurried conversations, stopping when Hugo & I appeared. Delegations & despair. What a way to be greeted."

So she sat everyone down, and one by one they worked through the problems. As for the baboons: no more stones thrown. As for the bananas: Jane thought, and everyone else agreed, that it was best to cut the provisioning schedule down to once every five or six days. They would still have some banana days in order to keep up the general observations, but the radically minimized schedule ought to discourage baboons from lurking about. And since the reduced provisioning would reduce the need for observing in camp, Jane went on to agree with the idea raised by Geza, Ruth, and Carole: that people might follow chimps out of camp and into the forest on the nonbanana days as an alternative form of general observations.

Following chimps! It was an exciting thought, made especially compelling for Jane by the memory of what she had done in her first couple of years there, when she was sometimes blissfully alone and often working by herself. True, when she first came, the chimps were extremely skittish and elusive, actually terrified of her: the two-legged alien invader suddenly dropped down from the moon. So her initial reaction to the idea of following chimps included that concern. She said something like, as Carole would later recall: "They always got uneasy if I tried to follow them, and I had to wait for them to move to another place and then approach them an hour or two later." But Carole, Ruth, and Geza were prepared for that objection. They had already tried following chimps—at least briefly, experimentally—and so they were able to say, "We've begun to do it, and it works." That finally persuaded her, and she was, in the end, delighted by the new plan.

. . .

Jane, Hugo, and Grub moved into the hut on the beach vacated by Tim and Bonnie, who moved into a new aluminum rondavel that had been recently erected near the beach. The staff had, in preparation for the van Lawick family's arrival, expanded the beach hut by putting in an open, roofed patio and then wrapping a heavy-gauge wire mesh around the whole thing, making it a baboon- and chimp-proof cage. That would prevent the baboons from stealing food; but much more importantly it would make sure that some eager chimp did not, after mistaking baby Grub for a baby baboon, carry him away as an interesting piece of meat. In any case, Jane liked what had been done to the beach hut, and she began calling it "the Cage."

Originally, she had dreaded living on the beach, thinking that it would stink of fish and be oppressively hot in the dry season, but it didn't and wasn't. It was perfectly cool, almost enough to require a sweater. And Grub was happy on the beach. He took walks every morning and evening when it was cool. He went for swims with his red water wings and then boat rides in his yellow wading pool. He was fascinated by the baboons, who would parade along the beach and show up at the Cage around lunchtime. And he loved the stray kitten taken in by Tim and Bonnie—descended from a cat owned by Iddi Matata of the fishermen. The kitten liked to stalk and attack Grub's toys, and that always made him explode with laughter. Jane hung a *kikapu*, a woven basket, on ropes suspended from the roof of the Cage. It made a lovely swing, in which Grub would doze while being gently pushed back and forth.

In the mornings, someone from the staff or one of the recent FWI volunteers would do the gentle pushing, thereby freeing Jane to go visit the chimps upstairs or attend to her correspondence and writing. Grub rarely cried when she left, and he was getting cleverer with each passing day. He could distinguish Mummy's shoes from Daddy's. He mastered building a stack four tins high. He would pretend to drink invisible cups of tea poured from an invisible pot. He knew what a horse was from his alphabet book, and he was able to point out, in a magazine, a wooden horse as another horse. Jane consulted Arnold Gesell's *Atlas of Infant Behavior*, which indicated that Grub was physically like a two-year-old, even a two-and-a-half-year-old, and in that sense he was flourishing. Gesell said that infants who failed to enter "the jargon stage" were likely to show "super-intelligence" in later months. Grub had not entered the jargon stage. In fact, he refused to talk altogether, even though he obviously understood many words. Jane intended to make a full inventory of his accomplishments as soon as he reached eighteen months. For now, though, he wasn't talking, and that remained a matter of concern.

. . .

Another concern. The Gombe Stream Chimpanzee Reserve was in the process of becoming a national park. Its newly elevated status would mean, among other things, that tourists would start to visit, and the vision of visiting tourists was actually troubling. The attraction of Gombe for tourists would always be the chimpanzees, but bringing in large groups of human apes to gawk at the nonhuman ones would obviously upset the chimps and disrupt the research. Also, since every disease that can infect people will also infect chimps, an endless stream of

visitors would recklessly endanger them. Those twin worries were serious enough to inspire several discussions between Jane and Hugo and
some high-ranking people in the Tanzanian Parks Department.

One such discussion brought into camp John Owen, chief warden of
the Southern Parks Department, and Steve Stephenson, chief warden
for Gombe, and their return to Kigoma afterward clarified another
potential worry about bringing tourists to Gombe. Nic Pickford was in
charge of ferrying the two administrators—along with two others, a
park ranger and the boatman Mpofu—back to Kigoma after the discussion. Since the Boston Whaler was by then damaged and awaiting
repairs, the five men stepped into a smaller aluminum boat. Given
that the usual outboard was out for repairs, the boat was powered by a
smaller outboard borrowed from Ramji Dharsi, Jane's friend and
favorite merchant in Kigoma.

Putting that much flesh into so diminutive a vessel was a bad idea.
Lake Tanganyika, 420 miles long and close to a mile deep in places, is
African's largest and deepest lake. In a body of water that more resem-
bles a sea than a lake, fierce squalls can materialize quickly and threaten
chaos and death—which is what happened to the five grown men in the
one small boat. Suddenly swamped by high waves, the aluminum tub
began to sink, at which point John Owen announced that he couldn't
swim. Nic had the presence of mind to dive in and unbolt the motor,
hoping to lighten the boat's weight. The unbolted outboard slowly
turned and slipped through a churning boil into the gathering blue
below, but even without that extra weight the buoyancy tanks were
insufficient to keep the boat afloat.

A steady wind continually pushed them—five men desperately tread-
ing water—away from land, although a couple of empty jerricans from
the boat were drafted to serve as floats and soon were supporting John
Owen. And by a random turn of good fortune, someone in a fishing village along the shore happened to notice their trouble, and two men
from the village quickly set off in a pirogue and were paddling to the
rescue. It took a half hour for the paddlers to reach the unboated men,
and the pirogue was too small to hold five additional bodies. Nevertheless, it had space for John Owen and one other, and the remaining
three—the stronger swimmers—gripped the sides as the pirogue was
paddled back to shore. "It was a very dicey 'do,'" Owen remarked after
it was over. "We were damn lucky."

That accident probably encouraged the two administrators to consider using the Parks Department's trimaran, the *Triton*—recently

acquired for the express purpose of shuttling tourists between Kigoma and Gombe—next time. Indeed, not so long after that incident, another well-placed parks bureaucrat showed up at Gombe in the *Triton,* bringing along his wife and mother-in-law for a taste of what life in paradise might look like once it became a national park. At the same time, they gave Jane a foretaste of what visiting tourists could look like. The mother-in-law behaved well, Jane noted in a letter home that July, but the administrator and his wife "marched in as though the whole place was theirs now it's a Park." They wandered among the chimps casually, even as researchers were trying to carry out their observations. The visitors stayed for two days, and how the Gombe researchers would ever cope with tourists showing up whenever they felt like it Jane had no idea.

There was one possible solution, but it would take a long time. Two years at least. Still, it was the only solution she could imagine: to divert the tourists to another part of the park some distance away from the research camp in Kakombe Valley. That way, the tourists would avoid disrupting the research, although of course they would still want to see chimps, who would have to be new ones who weren't part of the research group. This idea created the additional problem of finding those new chimps and then getting them used to being stared at by people. It was possible. Jane had demonstrated that it could be done. But it was also difficult and would require a special person to do it, and who might that person be?

2.

In early July, three exalted guests arrived, and the whole place seemed to vibrate with the excitement generated by their arrival. Jane was an enthusiastic hostess, eager to please, and pleased to introduce the chimps and show off the operation. *Exalted guests,* was Margaret's phrase, and I believe she spiced it with a dash of irony. But to Jane those three, and especially the latter two, were exalted without irony. They were important. They were distinguished scientists and professors, and they promised future funding and useful associations for Gombe as a major scientific research station. And since she and Hugo were becoming increasingly concerned about the state of Gombe's finances, it was possible to imagine that the future of the place and the research, even the chimps themselves, might be in the hands of those three.

Professor Phyllis Jay, a primatologist and anthropologist from the University of California, Berkeley, came to check on Tim and Bonnie's baboon work. Perhaps, Jane thought, there would be more funds for more baboon researchers from that direction.

Professor David Hamburg, a rising star in the psychiatry department of Stanford University, was beginning to look for long-term income from the United States that could support a select cadre of undergraduate students coming to Gombe from Stanford.

Professor Robert Hinde, a distinguished ethologist from Cambridge University in England, was promising to bring in British grants to support British graduate students and researchers.

Jane had known Robert the longest: ever since the final days of 1961, when she first arrived in Cambridge to begin her own doctoral studies and he became her advisor. He was a tall, handsome, blue-eyed, white-haired magus, a proponent and practitioner of a science of animal behavior that was designed to tame the evanescent truth of things through precision and rigor. Precise observations were important. Rigorous representation of those observations would become feasible by identifying as much as possible quantitatively, through numbers, which could be analyzed via stats and math. Jane was more the narrative sort, but her narrative style of data collecting—first one thing happened, then another—meant that detail camouflaged pattern. Robert wanted something simpler and more precise. Indeed, he thought, the data might become standardized and refined enough that it could be digested by a computer, which was why he brought with him that July a device for punching data into computer-readable punch cards. He showed it to all of the researchers one day and said they were going to give up using tape recorders and written notes.

Geza's first impression of the device was intensely negative. It was, he thought dismissively, a *gizmo*. A box with a strap and some keys. You were supposed to hang the strap around your neck, carry it about with you, stick a card in it, and push certain keys based on the particular behaviors you were watching. Robert had an educated overlay of good manners and pleasant behavior, but he was also determined about this. After he described how to operate the box, though, Geza said exactly what was on his mind: "Are you kidding? I'm never going to touch that box. You can use it all you want, but don't make me use it."

Robert said something to the effect that Geza should try it.

Geza said, "No. You try it."

Robert agreed that he would. The next morning he was up early with everyone else, and he walked out of camp following one of the chimp females. He disappeared for about three hours. He came back around lunchtime "a complete mess," in Geza's assessment. His shirt was torn. Blood was dripping off his arms and legs from all the thorny vines he had crawled through. The strap on the box was broken, and so he was holding it under his arm as he walked into camp. He had the grace to come up to Geza and say, "You're right."

Geza was right about the machine, which was too clumsy to use with wild chimps, but what about the method? This was really a conflict of style or approach, a philosophical struggle over how best to represent reality for scientific purposes. Jane briefly referred to that struggle in a July 16 letter to the family in England. "It was funny," she wrote. "Robert wanted lots and lots of information left out of the records because of the pile of notes building up. I agreed with lots, but we fought over lots too." Robert wanted her to use more charts, but she said she could never be sure that people had represented things accurately in charts—although she agreed to try. "And now, two days later, when he has seen his scheme in operation, he has agreed that I was right! Well, not quite so clear cut, but gradually things have been changed, and now it is quite super."

. . .

Yes, Jane was feeling very positive. The new provisioning system was well established. They were setting out bananas one day in every five or six, and she thought that was working well. With fewer bananas and fewer banana days, the chimps were coming in less frequently yet still often enough for general observations to be useful. And the baboons were much less a problem. She declared herself "happy about it for the first time since it started!" The general observation records were thus well taken care of, while people's follows of the chimps were starting to show promising results. Everyone who did it that summer—that is to say, Geza, Ruth, and Carole—loved following the chimps, and they had established a reasonable routine for doing it. Once every three days, each of them would select a chimp, a target individual, and follow that individual out of camp for as long as possible, taking detailed and timed notes as they went. "It really will be super," Jane wrote.

Robert had also helped simplify and clarify the in-camp note-taking system, making the notes more accessible by marking them with brief symbols at the margins representing twenty-two behavioral categories: S for submission, Gt for greeting, P for play, Th for throwing. That sort of thing. Doing this would make it possible for someone to scan the pages and more readily extract information on any of the twenty-two categories. In addition, Robert persuaded Jane that they should formally

Carole, Jane, and Geza lead the debate on bananas and record keeping.

separate the records for observations done at the banana-provisioning area from those records made by researchers following chimps out of camp and into the forest. These were both still essentially narrative records, but they were now officially divided in two: with general records done in camp to be known as A-record and everything done on follows out of camp to be known as B-record.

Aside from making such significant changes in the record-keeping system, Robert had agreed to help find an official senior scientist for Gombe, someone with the requisite credentials to supervise the work of all the younger researchers. Getting such a person would, according to Jane's letter home, be "very super indeed, and will take a huge load of responsibility from Hugo's and my shoulders—with Grub, and film editing, and book writing, etc., it is really too much to be solely responsible for a place that has grown so large and so fast as this." Robert was, furthermore, planning to send out, in around six months, a talented graduate student from Cambridge to study Gombe's leaf-eating monkeys. And he had recently arranged for Patrick McGinnis to become a graduate student at Cambridge, making it possible for him to develop and support his emerging interest in the sexual behavior of chimps. So Pat would soon leave Gombe and go to Cambridge for six months, then return for two years of field research, spend an additional year back in

Cambridge writing up his results, and finally be anointed a doctor of philosophy.

One problem: Patrick's draft board in San Diego had begun to send threatening letters. But perhaps someone could persuade the board to give him additional time to complete his research and education. That was the hope, and Jane wrote a strong letter on his behalf.

Altogether, Robert Hinde's visit to Gombe that July had turned into a "huge enormous success." Everyone was won over, Jane thought. Nic liked him and Margaret considered him a "sweet poppet." Geza was "thrilled with his intellect," as was Patrick. And Robert himself "adores" Grub and had "quite fallen for Geza's Ruth—who is willowy, quiet, and intelligent."

3.

Two or three short-term FWI volunteers that summer took over many of the basic chores, including typing, which gave Geza, Ruth, and Carole plenty of time to follow chimps out of camp for the B-record, and the first thing they discovered about following chimps seriously—no longer casually or experimentally—was how hard it could be.

Ruth wrote about the difficulties of chimp following that summer in regular letters home to her parents back in the States. She stressed the pleasures of such exploring and experimenting as well—for example, her discovery of a "really nice hill" covered with a soft, almost ethereal kind of tall grass that was wonderful to experience. On the other hand, when you entered a stretch of grass that was thick and coarse, sometimes sharp-edged, and around ten or twelve feet high, you could quickly become disoriented and begin to feel scared. It could be especially scary when you started to think about buffalo, who were known to lurk in those great stands of grass. And snakes. And yet, Ruth hastened to reassure her parents, everyone carried a snakebite kit and, in case of getting lost, matches and signal flares.

In fact, she went on in the same letter, there was actually nothing to fear about walking alone in the Gombe forest. Yes, there were animals all around, but they were hidden and rarely showed themselves. Indeed, the forest was nothing like she had imagined. She had originally thought Gombe would be like a very large zoo, but it was far more subtle and wonderful than that. "It's all so much more natural and really a great feeling when you feel . . . part of it all. Sometimes on my walks butterflies alight on me!" It was true that the hot climate and endless insects could make things unpleasant—and then there were the vegetative obstacles to consider. One time she entered a place where the grass was thick and high, and she became "scared of buffalo and thought I would nearly die of heat and thirst as it was almost 115. I could see and hear nothing at all and my throat was so dry that it ached and my mouth stuck together. Hundreds of little flies were buzzing in my ears and in my eyes." In steep places, she discovered, patches of tall, bent-down grass could become treacherously slippery. Elsewhere—for instance, inside some of the deep pockets and ravines—dense and thorny thickets became nearly impenetrable barriers for anyone who was human. "To get through such vegetation," Ruth wrote, "one has to assume all sorts of contorted positions and you come back aching from head to foot" as well as "all scratched and bloody, full of thorns and splinters."

Another time, determined to pass through a dense thicket, she followed an animal path that turned into a trail that turned into a tunnel, which she began crawling through until she was enveloped in vines and bush, hardly able to move. But she pressed herself forward. After what seemed like an hour of crawling she began to feel disoriented and then shaky. She cracked her head on a sharp branch, saw stars, started to bleed, then passed out. Or at least thought she had. When she looked up, an enormous baboon was standing less than a yard away and looking at her. He seemed monstrous. He raised his eyebrows, lowered his white eyelids, yawned wide open in a display of those great teeth; and Ruth could only respond weakly with strange noises and facial expressions. Then the baboon turned and ran off, while Ruth remained there with the sticky warmth of blood dribbling down her face, deciding at that point that "the B-record was certainly not worth this." She turned about and went back through the same tunnel: "What a joy to finally emerge into sunlight and open country where I could stand after 1 1/2 hours of crawling on my belly."

But the challenge of vegetative obstacles was multiplied many times over by that of the topographical ones, which Ruth described in a letter written after a day spent searching for some of the chimpanzee males. "Just yesterday," she began, "I was sent out on an expedition." Most of the males had been gone for a week or more, possibly because some wild fruit was coming into season. People were hearing chimp calls from a valley three or four miles to the north, and Ruth went looking for them. "This may sound simple," she wrote, "but the valleys here are huge and it isn't just a simple case of ridge, valley, ridge, valley." Instead, the ridges "branch in all crazy directions and it's quite complicated. I had quite a time as the country was foreign to me." She spent the entire morning at the top of a high ridge, where from an open vantage point she was able to look over into two great valleys—and still she saw nothing of the chimps.

Adapting to the heat, insects, vegetation, and topography was going to be an extended process: that much was clear. But more immediately and practically, Ruth went to Kigoma and bought a new pair of sneakers. And she learned to climb uphill and sometimes to crawl downhill on all fours like a chimp, although the chimps always did it much faster and more efficiently.

. . .

The A-record observations in camp were based on watching chimps act on an artificial, human-created stage in front of a human audience. Everyone who did the B-record follows that summer—in other words, Geza, Ruth, and Carole—recognized that in leaving the human stage and the human audience, they were entering the chimpanzee world and beginning to consider or imagine what it might be like to be an ape, a nonhuman ape, in an ape's world. Over time and as a result of their experience in B-record follows, all three completely changed their feelings about the chimpanzees and their ideas about what or who they were. It didn't happen for each of them in the same way or at the same time or on the same schedule. Having come to Gombe a few months earlier than Geza and Ruth, Carole had the most experience, and perhaps as a result she did the best and longest follows that summer. Geza, although he was physically stronger and more able than Ruth, was also more conservative and therefore slower to change his thinking about the experience and the chimps. But for all three the same thing happened: the chimps became somehow more real. They became creatures with worlds inside them, interior universes of experience, memory, and meaning.

Even Geza, who held on the longest to what he considered to be the proper scientific stance of cool objectivity, had experiences that summer that made him look at chimpanzees more sympathetically and with a greater uncertainty about whether he was able to process fully what was actually going on. For a week at the start of August, a bad case of malaria kept him in bed, occasionally delirious, and regularly oscillating between sweats and chills beneath a heavy pile of blankets. And it was after that bout of illness, at a time when he was perhaps still a bit dreamy in the head and weak in the body, that he followed a particular chimp out of camp and up along the ridge trail above Mkenke Valley.

The individual he followed had joined a larger group, so Geza spent some time moving with a group of several animals until, in the higher reaches of Mkenke, they all just vanished. Still hoping to find his target individual once again, Geza continued on until he was trudging up a steep slope out of the forest and into a zone of high grass, cooling breeze, and the occasional tree.

He was soaked and dripping with sweat. His arms were scratched, his clothes torn. His muscles were stiff and sore. He recognized that it was time to quit and return to camp. But first, he thought, he would catch his breath. Arriving on a high grassy knoll, he sat at the base of a stunted tree and paused to take in the view. Before him, as the land dropped precipitously, the grassy zone turned into forest and then dropped down to Lake Tanganyika. He could see, looking west across a few miles of glimmering lake, the rising palisades of a distant land: the Congo. The sky held a world of weather. There were clear openings and shafts of light amid great gray clouds and traveling dark storms sporadically lit up from the inside by shivering filaments of ionic discharge. The sun slipped lower and began spilling color into the sky and clouds, a seeping palette of changing mixtures and gradations of reds, yellows, oranges. A bright gold. A purple bruise.

The rising breeze drew the sweat off his body, pleasantly cooling him down; and as it moved, it imparted motion to the high grass in the form of ripples and waves. The regularity of this wavering drew his attention to a violation of regularity, which was, he saw now, a temporary division in the grass, a moving furrow. Then he noticed another moving furrow at another spot in the grass, and, in order to get a better view, he stood up and watched the two furrows move toward each other. It was like watching the pressure waves made by two invisible ships headed disastrously on a collision course.

Then suddenly from one of the furrows a head and face emerged, followed by a pair of long arms. Geza heard a bright scream and a hoot, and then he saw a second face and head emerge from the other furrow in the grass. The two heads turned toward each other, whereupon two big chimpanzee males rose more fully out of the grass, yet still crouching. Geza gasped with the pleasure of recognition. It was Hugh! Then he looked at the second figure. It was Charlie! They were brothers.

The two chimpanzees rose fully upright on their hind legs. With the hair on their arms and legs and backs suddenly erect from excitement, they raced through the grass and converged in a slamming embrace, arms around torsos, mouths pressed to shoulders, and they bounced wildly in that embrace. After a time, they both stepped back, and, still standing fully upright and gazing into each other's faces, they clasped together their right hands, shook those clasped hands up and down, and patted each other on the back with their free left hands. It was, Geza thought, exactly like watching two old friends coming together on the streets of an American city and greeting each other vigorously after a long time apart.

Finally, after all that enthusiastic greeting, the two big males found a higher spot out of the tall grass where they sat down side by side and busied themselves in a session of friendly mutual grooming, each picking particles and parasites out of the other's arms and back and shoulders while repeatedly pausing to turn their eyes toward the drama of the setting sun. Geza likewise sat down and turned to enjoy the same glorious sunset. This had not been part of the original follow, of course, and Geza never considered entering it into the B-record. But even if he had thought to do so, how could he ever describe what he now believed to be the evanescent truth of things: that the two brothers Hugh and Charlie were just really glad to see each other, that they very much appreciated each other's presence and companionship, and that they were now simply enjoying the peace and cool of the evening and experiencing a quiet moment of aesthetic pleasure as the sun spilled out all those colors in the sky over the faraway mountains of the Congo?

4.

Nic was born in England, but at the end of the war, when he was only five years old, the whole family shipped out to Kenya. His father was a farmer, and they settled on a piece of farmland at Kitale, up on Mount Elgon just below the bluff of the mountain.

It happened that Nic's chum at school was Jonny Leakey, son of the famous Dr. Louis Leakey. At the end of their final year of school in 1957, Nic and Jonny celebrated their graduation by hitchhiking down to Olduvai, where Louis and Mary Leakey were digging as usual for stones and bones. Jane Goodall was there, along with another bright young woman named Gillian Trace. The two of them were helping the Leakeys, while Jonnie and Nic showed up and began spending most of their time looking for snakes. Nic had not been impressed by Jane, but perhaps that indifference was related to the fact that he had soon fallen in love with her companion, Gillian Trace. Gillian was lovely. She was a looker. Of course, she and Jane were in their twenties at the time, while Nic and Jonny were just schoolboys, so they didn't count. But Gillian was still a very nice girl.

After the summer ended, Louis Leakey phoned Nic up one day and said, "I have a possible job for you." Nic went down to the Leakey house and was hired by the photographer and filmmaker Des Bartlett. He worked for Des for about two and a half years, spending most of his time in the darkroom developing and printing pictures. It was great fun, and he met all sorts of people, including the young Dutch photographer Hugo van Lawick. Nic was at the Leakey house working with Des on the day Hugo got the phone call that told him he had been hired by *National Geographic* magazine.

Then one day a few years later, in the early months of 1968, Jane and Hugo, now married, turned up at Nic's dad's farm and asked Nic if he would like to work at Gombe as an administrator. Nic had been to South Africa and gotten married to Margaret by then, so she could come along as well and make her place as the administrator's assistant. Given that he had grown up in Kenya on a farm—with dairy cattle, sheep, and pigs—Nic knew his way around animals and machines. He could fix things. He also knew photography and darkroom work, and he was fluent in Swahili. He was a steady and practical sort: reliable, resourceful, and cheerful. And, with that dark mustache, trimmed goatee, and the curly hair, Nic was also youthfully handsome. That he was married to Margaret was a bonus, Jane thought, because wasn't it better to have a married couple who could keep each other sane and provide a model of domestic stability for the others?

. . .

Nic soon felt completely at home at Gombe that summer. Margaret found it harder to adjust. The weather was so hot it drained her energy, made her limp. She and Nic had moved into a tent pitched on a flat spot near the stream and above the beach a bit, and she did enjoy looking out from their tent over the lake. She especially liked it at night when the fishermen paddled out in their canoes with those brightly lit lanterns hung on the bows to seduce the silver sardines, the *dagaa*. The fishermen would fan out in a great circle, and that sight reminded her of the lighted streets of Johannesburg at night. She loved seeing the dreamy drama of the lake just beyond where they camped, even though a hot canvas tent was not the first thing she would have chosen to live in.

Nic was content, however, and he kept himself busy. He cleaned out the hut where the generator was. Nobody was living there at the time, and he discovered an old wooden crate. When he went to pick it up, it transubstantiated with a puff of dust. Termites had eaten the inside of the crate and left the outside just a fragile carapace. Inside were hundreds of rolls of film that someone from *National Geographic* had sent to Hugo years before, all 8 mm film in little cans. The termites had eaten the cardboard off, and now there was a pile of splinters and dust with cans of useless film in the middle.

Well, Nic thought, things were not taken care of that should have been. The little Japanese generator didn't work. He had to make it work so they could use the radio telephone again. The kitchen had fallen down, so he built a new kitchen. The boat, the Boston Whaler, had been dragged up on the rocky beach too many times, and so the fiberglass bottom was ripped open with the polystyrene foam inside hanging out. He pulled the boat up onto the shore, flipped it over and soon enough had repaired it. After that they had a good boat to go back and forth in. Unlike Gombe's aluminum tub, which was now sitting somewhere at the bottom of the lake, the Boston Whaler wouldn't sink. You could cut it into pieces, and it still would not sink. He also kept the reserve's paperwork up to date and made sure the staff did their work properly. And then, of course, he was expected to do all the shopping.

Once a week he'd flag down a water taxi—or take the Boston Whaler, after it was fixed—and go into Kigoma to buy supplies. Stop at Ramji Dharsi's store on Lumumba Street, the only paved street in town, for the basics. Oil, sugar, flour, and such. Get batteries for the tape recorders and torches. And shoes. Tennis shoes made by the Bata company in Czechoslovakia were five shillings a pair. There was just a single thickness of canvas on top and the rubber sole on bottom, and they would last three, four, maybe five weeks, depending on how hard you used them, and then you'd toss them out and put on another pair. They were comfortable, and you needed footwear that was loose and cheap for ventilation but didn't keep long enough to begin rotting into your feet. Nic would also buy a week's supply of food.

He went to this Arab shop to buy meat. The back door opens. In comes a man with half a cow over his shoulder—they had small cows—slaps it down on the table, and it's still twitching. Just slaughtered and skinned out back. Nic would point and say: "Well, that bit." The butcher would hack off a piece, and Nic would bring it back to the cook at Gombe. Always with pawpaws, because pawpaw is a tenderizer. You couldn't hang meat for two weeks, like you do in a cooler climate, to mature it, make it edible. You would eat it the day you bought it, having tenderized it with papaya. But the meat was always rough, and the goat meat was rank, so pretty soon Nic decided to concentrate on buying fish, which was always fresh and good. And he would buy Johnny Walker Black, vodka, and so on. Then he would walk over to the Golden Lion Hotel, where the manager's wife sold delicious samosas as well as a special mango juice that mixed well with the vodka or just about any other kind of drink.

Nic liked to fix things, and he enjoyed going into town to shop, as well as helping people out when they needed help, but what he liked above all was catching snakes. That was his hobby and passion, something he had done all his life, and he was good at it. What was it about snakes? In the Bible, the Pharaoh's magicians turned sticks into snakes, and to Nic snakes were like that. They were so perfectly camouflaged, some of them, that all they had to do was stay still and you would believe you were looking at a stick surrounded by leaves. You could look right at a snake and not see it. They had this marvelous trick of invisibility, this astonishing, slippery sort of evanescence, and yet they were real and really there.

So he always kept an eye out for snakes, and in his free time he would try to catch some. He knew how to make a good snake catcher out of an old golf club with the head cut off. He would attach a bicycle brake handle at the top, run a cable down to a padded grabber at the working end. If he found a snake that was interesting, something unusual, he would catch it and put it inside a pillowcase, then knot the bag closed and hang it up inside the tent. That way he could keep a watch over the snake until he found someone headed for Nairobi who would deliver it to his friend Jimmy Ashe at the National Museum's snake park. So that was what Nic liked. Margaret, on the other hand, never really enjoyed coming into the tent and seeing a pillowcase with some deadly thing shifting about inside, and she developed the habit of having a quick look around whenever she went inside, just to make sure the snake in the pillowcase hadn't escaped to become a snake under the pillow.

5.

In order to rely less on meat and more on fish, Nic hired this fisherman, Alphonse. Nic paid him seven shillings a day, which was the equivalent of about a dollar a day, to give the cook first choice from his daily catch. Alphonse fished with a long net, forty or fifty feet long, with one side weighted down, the other side floating. He would pull that in, take out

Alphonse offers fish.

the fish, and the cook could come down and choose whatever fish he wanted, if they were having fish that day. If they weren't having fish, it didn't matter. The man still got his dollar a day, which was a good deal for him but also for the Gombe operation, because sometimes there were only four or five people in camp, sometimes a dozen or more, and they could feed the whole lot on fish.

Alphonse had lost a foot in a freak railroad accident, so he had only one foot and walked with the assistance of a stick. The other foot was just a stump at the bottom. But he was a nice guy, Alphonse was, and every so often he'd call Nic down to the beach: "Come see what I've caught!" Sometimes it would be a Storm's water cobra, a big one, and

Nic would put it into a preservative and keep it. One day Alphonse caught a *Polypterous,* which is an ancient fish, one of the things you see in fossils but seldom in the flesh. A *Polypterous!* It had lungs instead of gills and about nine flaglike fins on its back. Another time when Alphonse called Nic, he went down and the fisherman's net was up out of the water. It was jumping.

"What's going on?" Nic said.

Oh, it was bad. An electric catfish was in there, and it was making all those other fish just leap and leap and leap. It looked like the fish in the net had gone crazy or were under a spell, as if they were bewitched.

. . .

Another kind of bewitchery happened around the end of July: the beach baboons began to die. Jane and Hugo discovered the first dead baboon while they were walking along the beach and thought little of it. But then, over the next couple of days, they discovered three more carcasses, plus traces of two others. They began to suspect poison, but they were also worried that the baboons might have been dying from an epidemic of some obscure pathogen that could spread to the chimps. Hugo got on the radiotelephone to Nairobi and arranged for a veterinary chemist, Dr. Monks, to fly out the next day. He arrived and immediately went down to pry open the freshest carcass.

Dr. Monks thought it likely that the baboon had been poisoned, but, presuming poison had been used, he was mystified about its origin and shaken by its efficacy. The animal had sickened and died quickly. There was still food in the esophagus, which meant he had been feeling well enough to eat normally just before he felt bad and died. Monks examined the gut further and found minimal irritation. Most irritant poisons, he told Jane, leave a trace of irritation from the stomach all the way down to the large intestine, but in this instance the signs of irritation stopped within two inches of the small intestine. So the poison was impressive. It wasn't clear that Dr. Monks would ever be able to identify exactly what kind of poison it was, he said, since he believed it would be a substance unfamiliar to Europeans. But he swabbed some samples onto some slides, which he then packed up and took back to Nairobi.

Tim and Bonnie had left for a safari holiday during that same period, so Tim wasn't around to witness the horror of his research subjects dying. He also missed some of the interesting social sequelae, since the deaths of a few males led to serious conflicts and some terrific fights among the surviving ones, as they sorted out a new equilibrium in dominance relationships. The females and young, meanwhile, avoided most of the chaos by running into the lake at critical moments.

When he did get back and found out what had happened, Tim decided that it was his fault. It seemed likely that one or some of the fishermen had gotten poison from a traditional healer, a *mganga*, or perhaps someone skilled in sorcery, an *uchawai*. Either one would know about poisons, and the village of Mwamgongo, just north of the Gombe boundary, was big enough to have a number of healers, sorcerers, and even possibly a witch or two, although the latter, possessing inherent power, did not ordinarily rely on physical substances. The poison, it appeared, had been left out to drive off or punish the baboons. Or possibly to drive off or punish the *mzungu*, the white person: the one with the beard who walked among the baboons. The baboons had been eating the fishermen's sun-drying dagaa, after all, and Tim had made the fishermen stop chasing the baboons away from their dagaa as they normally did. Tim had done that by speaking to Iddi Matata, the sturdy old man with the white beard who acted as the fishermen's contact and negotiator. So the whole event was complicated and upsetting for Tim. First because of the baboons dying. Second because he had done the wrong thing by the fishermen, who were only protecting their livelihood.

. . .

The fishermen lived their lives by the moon. The moon went through its faces—round, side, coming, going—and sometimes it hid behind a veil or dripped with honey. Sometimes it was not a face but a milky breast or sleepy eye. The fishermen slept on the beach in huts woven from reeds, with reed mats inside, so they would not need to sleep on the sand or the pebbles. The days could be slow and hot, though, and they might feel weak and bored from the day. When the moon was whole and bright, they weren't even there. They were gone, having followed the Watu path—the People path—through the forest all the way up to the top of the escarpment and their *shambas* on the far side. That was where they lived and tended their gardens when they were waiting for the moon to change. But when the moon was turning away or dying and they were camping in their huts down on the beach, then, as the darkness of evening arrived, they came alive.

They climbed into their pirogues. They lit their lanterns, turned them up bright, and paddled away from the land to a place where they wavered like gods in the sky of the dagaa. They offered the dagaa a shiny promise of light, and the dagaa, stupid fish that they were, rose to it. Tricked like that by the promise of something good, they became tangled in something bad that pulled them shivering out of their own medium and into another: a terrible place of air and death. By day they were toasted in the sun and transformed into dry silver for the market. Too bad for the dagaa!

Most of the fishermen were Waha, but there were Wabembe, Wabara, and Wagoma, too, people who once came from the other side of the lake and now lived in their villages on this side. And there were some Waha fishermen who came from places other than Mwamgongo. They also came from Bubango and Mgaraganza, and so on. The fathers of many of them, possibly most, had been fishermen, and their grandfathers, too, and they were Waha.

Probably some of the old men could remember when the British made them stop cutting trees for the boat fires. The only way to stop cutting trees for fire—they had to light up their boats for the dagaa somehow, didn't they?—was to buy lanterns from the Brits. Tilly lanterns they were called, and they were better than burning fires on the boat. The lanterns didn't die down as fast as open fires, and they could be made very bright. But wood used to be free. Lanterns cost money, and they needed to be fed with oil, which also cost money.

Then the Brits put up new signs and made new rules, and the forest beside the lake became the Gombe Stream Chimpanzee Reserve. The Germans had done that first, true, and the Germans had called it the same thing, only they called it that in German. Now it was in the English language instead of the German one. But the Germans and the Brits: the only reason they could do any of that was that *the People had already done it for them*. What did they think? Did they really believe there were no animals—no leopards, no snakes, no screaming *sokwe* or scampering monkeys or singing birds—before they, the Germans and the British, came? No wild forest? Of course there was, and there had always been, and that's what everyone knew. The wild forest beside the lake was a sacred place where a person would not go, so when you cut a tree for your fire, for example, you didn't go into the forest. All right: where it was not so wild you might wish to plant something, such as oil palm. And you could cut wood that had fallen out of it, like a leaf dropped from a tree, or wood that was growing up fast at the edge of it like a weed. But the dark and tangly part, the wild forest part, was were where you did not go because, for one thing, that's where the earth spirits lived.

There were different kinds, but the Bisigo were the bad ones. Bad-hearted. Always dangerous. Some were active and had no home, always on the move, and when they traveled you might see them moving far off over the lake disguised as a whirlwind or a storm cloud or maybe a shimmering eye-bend in the distance. Or you might see them curling like fog on a still, cool, early morning when the fog rises fine and twirling in knotty vortices. But then again, someone might see them inside the wild forest, because the Bisigo could disguise themselves as pythons or spitting cobras. You could tell them apart from real pythons or real spitting cobras because they would turn invisible in an instant, whenever they chose. But not all the Bisigo were homeless. Some liked to settle down in special places. A special rock. A sacred pond or river. A waterfall. Or maybe a sweet-smelling tree, such as a mango or a kapok. So the inside of a wild forest like Gombe, with its tangled places and singing deeps where the earth is split open, could make the best kind of home for Bisigo to settle in. And since Bisigo might kill people or drive them crazy or make them speechless, the forest was just a good place to avoid.

True, a person could pass through it, go from the bottom to the top as the fishermen did when they took the Watu path and walked up to the top of the escarpment to reach their shambas. But anyone who knew the ways of the Bisigo would not stupidly court danger by entering the secret parts of the wild forest.

. . .

In his time of pain, I will imagine, Alphonse remembered the lake and the fish and the forests, and maybe that was when he decided that he, too, would become a fisherman. He had once wanted another kind of life somewhere else, and so he had left Kalema, the village at the southern end of the lake where he was born. He was an Mfipa of the Ufipa region down there, but he came north and settled in Gingu, which is the village at the northern edge of Kigoma by the last stretch of the railway line as it enters town at Kinkundu Road.

Alphonse was young then. His arms were strong. His back. He had two feet. Then he stepped one way when he should have stepped another while a train was moving down the tracks. Maybe he was working, and maybe he was playing. Maybe he had been drinking too much, and maybe not. And maybe what he could best remember afterward was floating above his own self and seeing himself writhing like a fish does when you pull him out of water. Was that real? The pain was real, something you know is pain and not something else. Like torture that does not stop. And maybe that made him think of the Tortured Man and made him look up and consider the little tortured man above his head. Because there he was, the tortured mzungu hanging on a wall above his bed, a wooden or clay tortured man with his head turned to one side, ribs sticking out like veins, red blood dripping down a milky face, black things like thorns sticking out of his head, his arms spread out wide, skinny white legs crossed at the feet. And there were the *wazungu* ladies, too, because there Alphonse was: being served food and drink and washed and wiped and made to go to the bathroom by wazungu ladies, some pretty, some not so much, but all of them wearing white. White caps that looked like white birds sitting on their heads. White costumes that flapped around their bodies like restless wings.

Alphonse could still feel the foot. Sometimes it itched. Sometimes it wanted to wiggle, and he would think he could make it wiggle. But he couldn't see it because it was gone. Later on, when he was well enough to sit up and bend over and reach down to touch it, the foot was still gone.

The pain lasted a long time, and perhaps when it went away there came another kind of pain: from knowing that he was no longer attractive in the way he used to be. Alphonse had a triangular face, a good one with a firm chin, wide cheeks, bright eyes, bright even teeth, a ready smile, full shoulders, strong arms, narrow hips. Because he was an Mfipa he had that facial scarification, scars going way back toward the ears, which was distinctive and attractive. He looked good, and people say he was charming and witty. He may have had the self-confidence that comes from being a good-looking man and knowing it, and maybe he was smart enough not to know it too much. But all that was gone now, or seemed to be, and who could blame the ladies in town who turned aside and pressed their lips together as he hobbled down the street. Because it was not just beauty that was gone. It was pride, and it was the ability to work, to earn. What could someone, a man, do without two feet?

Fishing was as easy as any other thing he might have tried. He could put himself into the water, where he would float, and floating can give someone the feeling that he is whole again. In the water a person can go from one place to another without having to hobble. A person can feel confident in the water, not have to balance all the time on a stick. And balancing with a paddle that's shaped like a broad-headed spear can make a man look less like a cripple and more like a warrior of the lake going to do battle with the fish.

The big problem was standing upright in a pirogue, which was tippy. Also, the dagaa fishers used a *kahwensulo,* which is a net stretched into a big circle with a long handle. They dipped their kahwensulo into the water and scooped out the dagaa. But how can someone stand on one foot in a pirogue and swing a kahwensulo down and up? So instead of becoming a dagaa fisherman like the Waha, the men who gathered silver by night, Alphonse became a man who gathered gold during the day. That was the way the Wafipa did, and that was the way Alphonse, being an Mfipa himself, did now. He got himself a *makira,* which was a long, fine net that he could stretch out in the water. One side of the net was held up with floats made from a corky-bark tree, and the other side was held down with weights. He could leave the net stretched out and floating like that overnight; it was invisible, so the fish swam right into it and got caught by their gills. In the morning, Alphonse untangled the fish and tossed them into his pirogue. They would land there and flip until he struck them with the hard handle of his paddle or until they ran out of breath and died on their own. Being dropped into a hot boil of pain for a long time can do strange things to a person. One person boils hard. Another person boils soft. It may be that Alphonse boiled soft and, because of that, didn't like to see suffering in someone else. And when he pulled the fish out of the water, tossed them into the bottom of his pirogue, and watched them flip and writhe there, did he, in his mind, see himself lying on the ground flipping and writhing with a foot gone? Maybe that was one more kind of pain, or an itch, and maybe it made it hard for him to kill fish. Even fish!

People said that Alphonse was the first person to bring a makira that far north on the lake, and so, although he may have started out as a lost wanderer, over time he became a cultural pioneer of sorts who, unlike anyone else, could fish whenever he wanted to. The Waha could only fish when the moon was coming, going, or gone, but Alphonse would just set his makira in the evening no matter what the moon was up to and then return in the morning. He used his makira to catch fish bigger than the little silvery dagaa that the others were catching, and while it was true that the bigger fish were not always as easy to sell, they still had value.

Alphonse also, after a time, learned how to catch *singa:* great catfish weighing as much as sixty kilos who were a prize of the lake. They were special fish. People said that you should not have sex before fishing singa, and they said that when you did intend to catch one you could find him by listening. That's what Alphonse did. When the lake was still, he would dip the blade of his paddle into the water, put the top of the handle up to his ear, and listen. He could find singa that way, listening to their underwater sighs and groans, and sometimes he could tell which direction they were going. If he heard one leaving an area he would head that fish off in his pirogue and then slap his paddle onto the water to drive the fish back. That gave him time to set the net.

6.

Meanwhile, a faint but lingering miasma of bad feelings that summer had begun to test the social equilibrium of the wazungu. On the surface, these were cultural and lifestyle conflicts, with the fragrant clouds of marijuana smoke down by the beach being the most obvious locus of contention. The short-term volunteers from FWI in Nairobi set up their tents down there, and they brought in guests, other FWI students from Nairobi who came as casual visitors and hangers-on. One of the FWI visitors, one of the boys, had bought a pound of grass in Ujiji, the town just a few miles south of Kigoma, and it was a potent strain. Some of the kids down at the beach were getting royally stoned.

Tim figured that whatever people chose to consume for relaxation purposes was their business, certainly not that of people upstairs who were relaxing with their own tinkling glasses of C_2H_5OH.

Geza thought that the FWIers down at the beach were, as he once expressed the concept to me, a bunch of "spaced-out teenagers from a world based on freebies of everything: freebie sex, freebie this, freebie that, plus a lot of drugs."

Carole considered Geza to be rigid and self-righteous, and although he had a sharp mind and a college degree, he was bound in his own perceptual straitjacket. He did not love the chimps or understand them in the way she did. And maybe the sex that supposedly went on down at the beach was not as free or loose or wild as certain people upstairs thought. Anyway, what was Geza doing with those *Playboy* magazines that came in the mail every month?

From Margaret's perspective the reason no one could get along was that half the researchers were long-haired American hippies and the other half were regular Americans—different but just as hard to put up with.

For Nic, part of the problem was language. The FWI kids said what they wanted to say. One of them was a vegetarian who refused to eat the fish. Jane brought her vegetarian hamburgers. She made the effort to buy those things in Nairobi and bring them down. The girl announced out loud that the vegetarian hamburgers were "garbage." "Why do we have to eat this garbage?" she said. She was like that. If she didn't like the bread or didn't like somebody, she'd say so. There was no holding her back. It was part of the FWI language thing. The FWI kids were just very loose, Nic thought, in language and behavior and everything else. Cussing was part of the way they talked, and they didn't mind being seen naked. They'd strip their clothes off and go on the beach, into the lake. They were just kids who had been sent to this funny Quaker institution to get their degree in what? In life?

Jane hadn't known about the marijuana-smoking at first, since no one wanted to tell her. When she at last found out, she was very displeased. "Never any more hippies again," she vowed in a letter to the family in England. And yet, of course, the problem was deeper than that, given the compound ripples of discontent running this way and that across their isolated little community; and the isolation itself made all the possibly trivial irritations and petty grievances more serious than they would have otherwise been, since there was no easy escape from one another. They were like space travelers in that regard.

Jane wrote home later to say that she and Hugo were having "conference after conference to try and sort out people being able to live together." It was not easy. "But by getting everyone together, and bringing out personal grudges and grievances—as we did last night over dinner—we seem to have got people to at least think that they can talk to each other politely. And no more FWI's!!"

. . .

Then Dominic Bandora showed up one evening, walking along the beach from somewhere else until he reached the research camp. He had been the first cook at Gombe, having started with Jane and her mother when they arrived in July 1960. He had shown himself to be a skilled cook and a charming man on whom Jane depended a good deal during her first couple of years at Gombe.

Dominic seemed to tolerate the isolation at Gombe well enough. In fact, the isolation was probably good for him, since it tended to keep him at an inconvenient remove from alcohol. But there was plenty of alcohol in Kigoma or elsewhere along the shore, so he would occasionally go on a drinking tear until he was undone enough to reveal a second, less congenial personality that could be belligerent. He once pulled a knife on Jane's mother, who had to talk him into putting the knife down. He was finally fired from his job in April 1965, after twice leaving camp without notice when Jane wasn't around. At any rate, here he was now, being his better self, smiling and friendly and catching up with

Jane on family matters. His daughter, Ado, was, he told her, *mkubwa sana sasa*—"all grown up now." Then he asked for his old job back, and so Jane hired him again. Thus, young Sadiki Rukumata, who had been the cook for some time, was demoted to assistant cook, and the quality of the meals improved. At least Ruth thought so, noting the appearance of some excellent desserts—chocolate cakes and apple pies—and describing Dominic in a letter to her parents as "a small, funny old man, extremely proud and an excellent cook."

Dominic was an Mfipa like Alphonse, which meant that Alphonse now had a natural ally in camp, a brother in the African sense, someone from the same tribe and region and background. Alphonse was funny and fair, and he became friends with a number of others on the staff, but now Dominic made sure to walk down to the beach every morning and sort through Alphonse's daily catch. That was good, but then Dominic went off one night and came back drunk. He started arguing with the other men at the kitchen, and when Nic told him to leave the kitchen, he said, "That's my kitchen!" He wanted to fight. The following morning, of course, he was terribly apologetic. The rest of the staff: No one drank like Dominic. Rashidi Kikwale was a Muslim and didn't drink at all. Tall and lean, he was a reliable and forthright man from Mwamgongo. He had been one of the first people Jane hired—second only to Dominic—back in 1960, and then he had worked as a field assistant. Now he was head of staff, the senior man, the *mzee*.

Counting Rashidi, the African staff were about a half dozen strong during the summer of 1968. Hilali Matama, just hired that year, was training to become the first of the new field staff: assistant to the researchers going out in the forest. He was a small, serious man with a superb memory for details about people and chimps and a good sense of the landscape. Then there was Yusufu Mvruganyi. Yusufu was Tim's favorite, and, in fact, he was paid through Tim's grant from Berkeley. It might be accurate to describe him as Tim's personal assistant or servant, but he was about Tim's age, a little younger, and Tim came to trust him completely. He was just a great guy, Tim thought, who wanted to figure out how to advance his life, which probably meant getting more education.

But here was the thing: Gombe offered fair employment for someone lucky enough to get it and willing to live in the middle of nowhere. Many ambitious young Tanzanians believed, with good reason, that the way to advance was to move urban. Living in the bush was not for everyone. And yet a few good men got jobs at Gombe, brought their wives,

Dominic sorts through the fish.

raised their families there, and joined the community of the staff village located just above the beach and a little south of and downhill from the scattered constellation of tents and huts inhabited by the wazungu that summer.

The Africans provided the labor, the backup, the cooking and cleaning and organizing of things. They negotiated for and brought in boatloads of bananas from Mwamgongo; and after dinner was cooked, they hauled the food and dishes up to the dining house at ridge camp, then cleaned up after. They were the foundation of the entire operation. But when the day was done, and the cool evening air moved in, along with the echoing chorus of night insects and the rustle and cry of nocturnal life, the Africans went back to their families in the staff village. It is likely that, as they relaxed, they occasionally talked about the wazungu, but surely not for long. Meanwhile, the wazungu, gathering for dinner at day's end, relaxing up at the dinner pavilion in the same cooling air, becoming hypnotized by the same chorusing night insects and the same rustle and cry, would occasionally talk and sometimes argue about the Africans. Or, rather, they would sometimes argue about the way the Africans should be treated.

It was another locus of contention, actually. Nic, who grew up with African workers on the farm and African helpers in the house, thought there had been some good conversations on the issue. Still, the Americans, especially the FWI kids, just wanted to be chums with the Africans. They couldn't understand that the African staff were paid to be there and that you had to deal with them differently from your friends. Trying to treat them as friends, hoping they would think of you as a friend, was simply a form of social exploitation. They weren't paid for that.

7.

In her early days at Gombe, when she first started living in the ridge hut, Carole would bolt the door at night and then crawl under her mosquito net gratefully, knowing that the giant poisonous centipedes on the rafters weren't going to fall onto her, because the net was tucked entirely around her like a silken skin. She was afraid. But after a while she decided that she really liked living out by herself in the middle of the forest, and so she stopped bolting the door at night. After another while, she started leaving the door wide open. Normally the day began at dawn, and she would get up, put her clothes on, go upstairs. But when she didn't have to get up early as the first observer, she might linger in bed and occasionally hear chimps pass by or even see them.

One morning, she was startled awake by an anomalous sound close by. That initial startle was followed by a developing awareness that someone had crept inside her hut. Careful not to move, she slowly opened her eyes. Although she was terribly nearsighted without her glasses, she could see a chimp sitting on the floor next to her bed with what seemed like an expression of wonder on his face. It was Figan. Carole decided that he had been watching her sleep.

Of course, Figan knew Carole, but he had never seen her lying in bed sleeping like that, had never before been able to walk in and look. It was a new experience for him, Carole thought, and she believed she could

Figan.

see that in his eyes. She was enchanted. She said, "Hi Figan," and Figan stood up immediately. With an awkward style of body movement that seemed to express surprise and embarrassment, he went over and looked in the waste basket, as if to say, *Oh, this is not me. I wasn't watching you when you were asleep.*

. . .

The experience made Carole begin to think of Figan as a friend. Then came the Golden Summer and the B-record follows, when she got to know Figan and his older brother Faben a lot better. In fact, the follows utterly transformed her knowledge of the chimps and her feelings about them.

A number of things made the follows work well. One was that the followers—Carole, Ruth, and Geza—had already become adept at observing the chimps in camp. From their camp experience, they already knew the chimps so well that they could recognize an individual's voice from a distance. They could see someone from behind, or just an ear and part of the head, and often would know from such fragments the whole. Who it was. A second thing was that the chimps had become completely at ease with them. The chimps were used to certain people throwing stones in their direction and knew they were aimed not at them but at the baboons. The chimps knew those humans were theirs.

A third factor was the terminology: the researchers were recording with a clear idea of what they needed to say. From all that time spent working on general observations, they had become skilled at making particular verbal statements about what was going on, tracking the present moment with words. The words weren't as good as the reality, of course. With words you focused on the chimps' social behavior and, to some degree, how they were interacting with the environment, but it was never as if the words made the whole reality.

For Carole, that was a gift. Being a practiced observer freed her from her chattering mind, gave it a task and thereby kept it from taking over the entirety of her consciousness. She could semiautomatically mutter into the tape recorder: "Figan stands up. He sways from side to side. He begins to pant-hoot. Faben stands up. The two of them run together." And while she was muttering in that manner, she would be looking at all the other chimps in the group and noting how they interacted. And, too, she would be sensing the air, feeling it slip along and stroke her skin. She would be listening to the whispers of the wind and watching the leaves shiver and shine in their leaf dance. She absorbed it all in what she privately thought of as the Wonderful Wordless Witnessing.

One of her follows focused on Mike. Although Mike was her target individual, he was part of a larger group that included Goliath, Pallas, Pepe, and Sniff. At the start of the day, they went south from camp, climbing two-thirds of the way up Sleeping Buffalo before pausing on a ridge. It was always hard to keep up with chimps when they went up a steep slope. They climbed it nonchalantly, while Carole struggled after them, panting and puffing. She managed to keep them in sight, however, and once they sat down at the ridge, she was able to catch up and join them. From their bivouac at the ridge, they could look down at the camp, and so they watched the camp—spied on it—for over an hour. Carole felt a joy in being with the chimps and joining them as they spied on the humans in camp. They were waiting to see if any bananas were going to be put out. Before they left Kakombe Valley, they wanted to be certain there would be none, so they stayed there for a good hour, then finally got up and went on.

Whenever Carole was with a group of chimps like that and she had a chance to join them while they were resting, she would. Once they began moving again in single file, as they usually did, she would try to insert herself as second in line, right after the leader. Since she was the most inept of the group, one by one they'd pass her, so starting as second in line enabled her to stay with them longer than she would otherwise have done, before the last chimp disappeared from sight.

That day, in her follow of Mike, the group climbed to the top of Sleeping Buffalo and then left the camp valley, Kakombe, and turned into the next valley south, which was Mkenke. That's when Carole fell behind, with only Pepe still in sight. She had hoped that Pepe being part of the group would give her a better chance of keeping up, since he was crippled with a bad arm: one of the victims of the 1966 polio epidemic. He could still travel faster than Carole, but even after he, too, had moved out of sight, Carole thought she knew where the group might be headed. She continued through Mkenke by herself, climbing up the Mkenke Watu path—the People path on which the fishermen walked from the camps on the beach up to their shambas on the far side of the escarpment—and she found the chimps along the edge of the path, resting. They saw her coming. She was puffing, breathing hard, climbing the path that went from the forest into the higher grassland with the slippery grass. She struggled up to where they were settled down and resting, and as soon as she got there, the chimps just got up and moved on. *Oh, gosh!* she thought.

She continued after them, panting, completely out of breath, but now following a narrow path across open terrain. It wound down into gullies and back up again, but Carole didn't fall behind. She stayed with them. They'd pause and eat something here and there, then carry on. They scrambled down into a dry ravine and scuttled up onto a grassy ridge, and by early afternoon they had left upper Mkenke Valley and gone farther south, into upper Kahama Valley.

They traversed a gallery forest of Kahama in a flat basin. Because the trees were mature, not enough sun penetrated to support much undergrowth, and that made it open and easy for Carole to walk through. She was keeping her position as third in the line, with Mike and Goliath ahead of her, and Pallas, Pepe, and Sniff behind. Suddenly, a group of ten banded mongooses appeared, humping along in their short-legged, long-tailed, low-to-the-ground fashion. Carole had never before seen them up close like that, and she was astonished because the mongooses weren't afraid of chimpanzees and seemed to be mistaking her for one of the chimps. They began to filter through. Here were the chimps, walking single file in one direction, and the mongooses traveled as a bundled group and at an angle to the trail the chimps were on, so the mongooses filtered through the chimp line. As they did, one mongoose who was passing behind Goliath and in front of Carole, stopped and stood up at the sight of her. He looked right at her and made an alarmed squeaking noise, *ee ee ee ee ee ee,* as if to say, "What the hell are you?" Carole said, "Hello, mongoose."

It was amazing to have this other being, this member of another species, stop and recognize her existence, stand there and give his impression of her. After Carole greeted the mongoose, he or she dropped forward, paused a little longer to contemplate, perhaps, the oddity of a bipedal ape in a line of quadrupedal ones, then joined the rest of the mongooses.

The chimps crossed Kahama Stream several times that afternoon, circling around. Goliath had taken the lead, and they were sauntering along at a comfortable pace. But then Goliath stopped, paused, turned around, and came back to Mike with a big fear-grin on his face. Mike stepped forward, and the two of them looked forward and into the trees. What had Mike been afraid of? Possibly he imagined there was some kind of threat, but it turned out that Goliath had only been spooked by the rustling of monkeys. Carole could hear the monkeys ahead when the chimps became completely silent. Previously, they were crunching on the leaf litter, walking casually, and Carole's walking had not been that much louder than anyone else's. But now their movements became silent, and she was a clumsy human wearing shoes. Only by moving much more slowly could she be quieter, so she dropped back. They were stalking the monkeys, and she was afraid she would ruin the hunt by being a human there. Sure enough, soon she stepped on a twig that went *crack!* Pepe was the last chimp in line, Carole behind him, and he turned around and made a minor threat gesture—cocked wrist and sudden lift of the forearm—and a soft bark.

From where she was then, she could make out monkeys in the trees—either red tails or red colobus—and it appeared that the monkeys had been alerted. They began giving alarm calls, while the chimps stopped stalking and just looked at each other. Carole thought she had ruined the hunt, and it seemed to her that the chimps were puzzled by that, looking at each other as if to say: "Don't the monkeys know this is just a human? They're harmless."

. . .

Great things happened on the follows. Yes, the banana provisioning was still essential. Without it, you could never keep track of everybody because the chimps were always so scattered, but the follows were an entirely different and more intimate way of being with the chimps as well as being in the forest. Making the follows officially part of the data collection—the B-record part—turned out to be a good thing as well.

Sometimes the follows didn't continue very long, but a good short follow might go on for two and a half hours. Even six hours was considered short. And they were all wonderful. Carole loved them all. But the long follows, going from dawn to dusk, were the best of all, and by the end of the Golden Summer, Carole had demonstrated to Jane and Hugo that she could do them better than anyone else. She was big, strong, skilled, and determined. Jane and Hugo recognized that. You had to have skill and determination above all, since the chimps could always outwit you, and although they might slow down sometimes to accommodate you, they were not going to waste that much time in accommodating the two-legged aliens when there were so many more important things to think about.

. . .

One of Carole's favorite long follows focused on Figan, and during this follow she found that Faben was supporting Figan in his displays out in the forest. Faben's support made Figan stronger. So she came back to camp with this story of the two brothers, the older Faben supporting the younger Figan. Jane had known Faben before he had been crippled by polio, and she was convinced Faben would have eventually become the alpha male of the entire community: king of the realm.

That never happened because of the polio.

Faben had been fabulous when both his big arms were whole. He was confident and powerful. Magnificent, really. And Carole thought that he was pretty impressive in his adaptation to that dreadful polio paralysis. His huge right arm was almost totally paralyzed, and he adapted to it superbly, despite the limitations of the chimpanzee skeletal structure. You'd see a group of those apes traveling single file, down on all fours and walking on their knuckles, heads loose and relaxed,

Faben walks upright.

seeming only barely to be looking around. Moving that way, in their quadrupedal gait, they were much shorter than people. But then you'd see Faben come along in his waddling, bipedal gait, walking upright with his good arm holding his paralyzed right arm, which was this swinging dead weight. It swung, and he would steady it with the other arm. And with those two arms out of commission, he would often walk upright, and his head would be up. He would be looking at you in the way a human being looks at you, and he had the most beautiful face. He had a wide, classically beautiful ape face, one of the most beautiful faces of any of the chimps. He was a joy to behold. Yes, Faben was one Carole loved dearly.

Before being stricken by polio, Faben had always dominated Figan, who was the little brother. It was normal for an older sibling to push around the younger one. Then, when Faben became paralyzed and frightened by this curse that had taken over his body so mysteriously, Figan saw his chance. He had been saving up his grudges, and for many months he paid Faben back for all that domineering-older-brother business. Figan was perhaps not quite as noble or generous as Faben to begin with, and maybe he was meaner to Faben than Faben had been to him.

Carole began this long follow with the younger Figan as her target individual. He was the one she would focus on. So she started out behind Figan, who was traveling by himself. But then he met up with this big group, and Figan's crippled older brother, Faben, was part of it.

First they went north. Then they reversed, came back through Kakombe Valley, and went on to Sleeping Buffalo. The midday heat had arrived, so everyone took a break, sitting around lazily and casually grooming each other. When the group was relaxing thus, the ambitious upstart Figan decided he wanted to show what a big guy he was, so he stood up and began to pant-hoot, then went into one of his dramatic displays, running on all fours across the open summit: the big flat area of grass and scattered trees on top of Sleeping Buffalo. And when Figan began to display, Faben got up and ran along with him, hooting and moving upright on two legs beside Figan in a parallel fashion. It was a coordinated dance, a dramatic expression of power and prowess, and with the two brothers doing it in parallel, the whole thing was impressive indeed. Several chimps were forced to bow and scatter and squeak, submissively skulking out of the way.

The parallel display was a pattern Carole had never seen before, and when she got back to camp, Jane was interested because she had known the other side of things: when Figan was paying Faben back for being a typical older brother. But now, it was becoming clear, the two brothers had outgrown that. They had become friends, and crippled older Faben had decided he would support ambitious younger Figan. That's what Carole saw. She was the first observer to get a strong sense of this new development, and she was able to do that because she stayed with them the whole day. She watched how they were together, and in her B-record she began to take note of everything Faben and Figan did. That was part of the observation protocol for the B-record: You had your target individual, but you were also supposed to identify all the others who interacted with the target. There Faben was, backing up and supporting Figan the whole time, so Faben became relevant in the record.

It was a glorious day. She just loved it. She loved her subjects and the whole big group they were moving with, and when the light became angled and saffron, she watched them climb into the trees, bend down branches and weave the branches into platforms or nests, then lie down in them—and it seemed to Carole as if they were tired and felt content, grateful to be tucking themselves into the safe and squishy beds they had just made after a long day. That was the first time she felt clearly that she was experiencing the world with them inside their wilderness, rather than being in a research camp with a bunch of humans and all the human-based requirements for observations. The chimps had nested before it was totally dark, and in the fading light she headed down the Mkenke Watu path toward the beach. As she came down, she thought: *This is the best day I've ever spent in my life. There is nothing I would want changed! Nothing!*

She was seized by a sense of timelessness, and all her uncertainties and concerns and troubles dwindled into nothing and vanished. She thought: *This is reality. This is the most real experience I've ever had.* It became clear to her that the present moment, the very moment she was passing through, was, like an invisible page turning in an invisible book, the best of all that life could offer. She felt privileged to experience it so completely and to be so full of joy. That powerful sense of timelessness, she thought, had to do with the chimps. They were like gurus. They were wise beings always alive at full stretch, and they brought her into that same state of immediacy. The feeling was so different from her ordinary mental state, which was more like being at a party somewhere and looking surreptitiously at her watch, wondering whether she should be somewhere else or with someone else. Carole felt she usually dragged around a lot of mental baggage, while the chimps, although they were not always happy and not always in the best place, were always in the moment. So much of her thinking and the writing in her journals returned to those old themes of her own uncertainty and dissatisfaction and malaise, of wishing things were different, but here she was, walking down the trail and thinking: *I'm not wishing for anything to be different. This is good enough. This is it. This is fine. This is better than good enough. This is marvelous. It's just everything I've ever wanted.*

IV

Transitions

(September 1968 to March 1969)

1.

Early that September, Jane and Hugo asked Carole if she would be willing to take on the task of taming a group of stranger chimpanzees—ones seldom seen, never identified—living three valleys to the south, in Nyasanga Valley. No one wanted the research chimps overwhelmed or the research station in Kakombe Valley overrun with tourists, they said, which was why the Parks Department had agreed to establish a tourist station at Nyasanga. Tourists would be coming to see chimpanzees, certainly, and so the Nyasanga station would need some who had been tamed well enough to show up for bananas when bananas were offered and stay long enough to be seen when seeing was desired.

Carole loved the chimpanzees she already knew, the camp chimps at Kakombe, but she agreed with the principle. She, too, didn't want the research camp overrun with tourists or the chimpanzees there spoiled or destroyed. That idea was a horror to consider. At the same time, she really did not want to tame a bunch of stranger chimps way down in Nyasanga. At first she suggested a compromise, saying, weakly: "I'd rather stay in the main camp, then once in a while go down and look for strangers."

When she went on the long follows, she occasionally came to the edges of the research chimp community's territory and saw chimpanzees she didn't recognize, strangers who were not part of the research chimp community, and thus she had become interested in the still poorly

Carole follows as Hugo films.

understood issues of territoriality and the relationships between different communities. That interest may have been part of the reason why Jane and Hugo believed she should be the one to tame a new group. But more to the point, as Jane explained, "You're probably the only one who could do it, Carole."

The comment flattered and enticed her, and she began to imagine that she could accomplish something heroic, do something similar to what Jane had done back in 1960 when she first started the whole enterprise of watching wild chimpanzees. Soon Carole was excited enough with the idea that she agreed to try. Unfortunately, at the very time of this conversation, her body was once again starting to give out. She understood that her old illness and the fatigue associated with it were returning. She could never tell Jane how serious it was, but she said enough that Jane and Hugo agreed she should go on holiday, see a doctor, and get some rest, and then she could return in December to start the habituation project for stranger chimps in Nyasanga.

Carole's holiday began in the middle of September, when she took the train from Kigoma to Dar es Salaam to see a British doctor someone had recommended. The British doctor offered no good theory about why Carole was so weak and tired, and after she had been in Dar for a few days, someone recommended a clinic elsewhere in the country run by an Austrian physician. She traveled to that clinic, where the Austrian said, "You've got tropical fatigue." Carole had no idea what that meant, but he continued: "You went back to work too soon. Your liver hasn't recovered, and so you really need to rest."

She traveled on to Nairobi and proceeded to rest at the FWI head-quarters: a big stone house in the diplomatic section of town where about three dozen American kids were spread out, grouped into various rooms and sleeping in sleeping bags. The place had a large, smoky kitchen, and the students cooked their own meals and talked a lot as they did. Some of them were going to the Starlight Club, a downtown nightclub that had live bands and recorded music (South African dance music, American rhythm and blues and soul), where they made friends with an African band. A few of the more sexually liberated young women in the group had done that. Carole wasn't, and she didn't. Instead, she met a boy—a man, really—from the new FWI class. Since he was a hulking sort who rode a big motorcycle, she thought of him as her Gangster Boyfriend. Since she had already lost her virginity during a brief fumble under the stars that summer, she also thought of him as her Second Lover. He was interesting, and she loved hanging on for dear life as he roared through the Nairobi streets like a raider on a steel horse, clinging on the back of that throbbing machine in her role as the raider's woman. She liked being with someone big and tall, since she was big and tall, and she enjoyed his sense of drama. One of his memorable comments was: "What does it feel like to be a goddess?"

She did her best to sleep a lot and avoid alcohol and stress, while also enjoying the pleasures of life with the Gangster Boyfriend and mentally preparing herself for the coming December, when she would return to Gombe to start her new job taming stranger chimps in Nyasanga Valley.

Back at Gombe, meanwhile, and early in the morning of September 18, Hugo, Nic, and Margaret were taken in the Boston Whaler to Kigoma. At the harbor car park, Margaret and Nic climbed into the VW bus, Hugo into the Land Rover, and they began the long drive to Ngorongoro. Three days later, they were sitting in canvas camp chairs and sipping tea inside the crater, next to the Munge River and underneath the giant fig tree, which was just then filled with birds filling themselves with figs.

Hugo was still photographing hyenas, and he worked at night in the Land Rover. It was a holiday for Nic and Margaret, but they also wanted to help. Nic would drive. Margaret would sit next to him in front. And Hugo would stay in the back seat with his camera mounted on the door and pointed out the open window. They wore crash helmets and safety harnesses and drove without headlights, navigating in the watery moonlight while searching for a giggling pack of thuggish hyenas. Once they found the racing beasts, they simply followed in their wake. But it was cold at night, even colder than it might have been, because Hugo had removed all the windows; and Margaret could not imagine where Hugo got his stamina. They would have supper at the absurd hour of three in the afternoon and leave at around half past five to look for hyenas. At last, as she explained to the others, she had had enough of untamed nature and preferred to stay in camp. No one objected, but she still stayed awake until the early hours of the morning when the others returned so that she could huddle down with Nic.

Jane had remained for an extra week at Gombe mainly because of a previously planned visit by the director of the Amsterdam Zoo and his wife. At the end of their visit, a telegram addressed to Patrick McGinnis arrived, informing him that his stepfather had died. Pat, needing to get to Nairobi in order to fly back to America for the funeral was at the last minute squeezed, along with the zoo director and his wife, into the same chartered plane that on September 24 carried Jane and Grub from Kigoma to the Ngorongoro Crater. The pilot deposited the two of them at the Ngorongoro landing strip, then flew the other three to Nairobi.

2.

Once, during the summer, Geza heard one of the FWI short-term volunteers talk about her physical attraction for the male chimps. She wondered aloud: "Which one of them would make the best lover?" That sort of talk went way beyond what he found comfortable. It alarmed and disturbed him. He also listened to Carole speak about how she "loved" this or that chimp, in her emotional way, and he considered that to be one more indicator of Carole's flakiness, her undisciplined, uncritical subjectivity. Carole had demonstrated herself to be a good follower of chimps, but she wasn't a scientist. She didn't understand science, and she didn't understand objectivity as a professional stance.

After Carole left in September, however, he began to imagine he could understand what she meant. To begin with, he started to recognize that he had favorites, real favorites, among the chimps. Leakey, for example. And in Leakey's case at least, it seemed that he returned the sentiment. Geza gradually started thinking of Leakey as a *friend*, which

Geza and Leakey.

was the only word he could ever think of to describe the relationship in plain English. To make the English less plain: There was a positive emotional association between Leakey and him that extended in both directions. Leakey seemed to like Geza's company. He liked Leakey's company. Obviously, the relationship was mute. They never talked to each other, since chimpanzees don't talk, but that muteness began to seem like a minor obstacle to understanding.

Geza told me about a time when he followed Leakey out of camp in the middle of the day. Chimpanzees often took a siesta in the middle of the day, and when that happened, Geza would have to wait for them. It would get boring, because he couldn't nap. He had to stay awake and be near enough to watch them so that they wouldn't disappear at an inopportune moment, and thus he usually included in his field pack a paperback to read when the chimps were napping.

Leakey did the usual thing that day. He decided to take a nap after going foraging. This happened somewhere near the Dell in Kakombe Valley, and the chimp picked an open area that was relatively grassy and rocky. As Leakey began hunkering down, Geza sat down with his book and started to read. But he noticed vaguely, at the periphery of his vision, that Leakey was having trouble finding a place that felt comfortable. He'd try out a spot, scoot around restlessly, move over to another spot, and so on. After a while, Geza noticed that Leakey had come over to sit down right next to him, so close that the ape's arm hair was brushing the hair on his arm. Geza just sat there. He didn't dare move, and he didn't know what was going to happen. Leakey next rolled over onto his side right next to Geza, his feet down the slope a bit, his head resting on Geza's right foot. Geza wore tennis shoes, so the sensation of Leakey's resting head was clear and direct.

But Leakey still seemed restless. He moved a little until his head was off the foot again, and then, after a few seconds, he reached up, grabbed Geza's ankle, moved the tennis-shoed foot over a few inches, and then put his head down on it again, shifting about until he was satisfied. He went to sleep and began to snore. Geza was frozen to the spot, couldn't move—but he was astonished and loving the moment. To have someone from another species, a wild animal, lie down and use a part of your body as a pillow, then go to sleep and start snoring, exhibited a level of trust that went well beyond anything he had ever expected.

. . .

Worzle was a completely different story. Geza always thought of Worzle as an odd character. Even the other chimpanzees seemed to think that, and Worzle was on the receiving end of some strange prejudice coming from both his fellow chimps and the humans.

For starters, no one could figure out how old he was. He looked old. He had a wizened, geriatric look about him. Perhaps he was Leakey's brother, people said, but no one was sure whether he would be the younger or the older one. It's true that the two resembled each other to some degree, and Worzle always seemed to seek out Leakey. Both clues suggested a fraternal connection. By the fall of 1968, Worzle was also worn and thin, and he moved cautiously and hesitantly, reminding Geza of an anxious and fragile old man. His physical condition could have indicated an advanced age, or it could have been the result of malnutrition early in life. Perhaps he had been orphaned before he was fully able to fend for himself. In any case, he was also low in the male dominance hierarchy, and his caution and habitual submissiveness meant he spent a good deal of time watching others, always careful to anticipate the sturm und drang that high-ranking males routinely provoked. For the humans who watched all this, meanwhile, Worzle was someone to pity, since it often happened that young males on their way up, physically and socially, would pick on him as a way to demonstrate their own expanding participation in the adult male world, their own developing capacity for dramatizing power and creating chaos.

Worzle with the white sclera.

Another thing about Worzle: his eyes. They were extraordinary. Owing to what was probably a simple genetic anomaly, and unlike those of any of the other chimps at Gombe, Worzle's eyes were marked by white in the area outside the pupil and iris, the sclera. Humans normally have white sclera, and that single anomaly gave Worzle a hauntingly human look. His eyes, combined with his anxious personality, made him look like a thoughtful and possibly fearful old man who was always looking about, always thinking, always concerned about what could happen next.

Given Worzle's old-mannish frailty, one might imagine that the researchers could more easily follow him, but that was not the case. His normal reaction to being followed was to vanish as soon as the opportunity presented itself. He was shy around other chimps but equally so with humans, and, among both species, he was shier with males than with females.

When he traveled with a group, it was easy enough for a person to follow him, since you only had to keep track of the rest of the group. But when he moved alone, following him was not so easy. Geza would usually stay some distance behind in order to reduce Worzle's discomfort, but of course staying too far behind meant Geza would soon lose sight of him, and then it would be a matter of seconds before the old chimp would do something tricky. He would veer off a trail, insert himself into a bank of lush vegetation, or mysteriously vanish in a flicker of light.

One time, Geza followed Worzle out of camp along the path that led to the Kakombe waterfall. Worzle was thirty feet in front of Geza, moving quickly, in his awkward, hunkered style, turning to look back frequently in order to check on Geza, scuttling along a little more quickly whenever it might have seemed that the human was gaining on him. They passed the Rock Pile and then a dry ravine. Worzle would stop from time to time, long enough to pluck off tender new leaves from plants along the way, stuffing them into his mouth to join the banana peels he was already working on, chewing away at the full wadge in the style of some person chewing a savory mouthful of tobacco.

They passed the Dell and advanced in the direction of the Kakombe waterfall, but just before they reached the falls, Worzle did his disappearing trick, cutting away and dropping through a leafy hatch into a dense thicket. Geza hurried up to the spot where Worzle had gone and discovered an animal tunnel within the thicket. Knowing that Worzle would just roll on through the tunnel, Geza nevertheless dropped down and began crawling through it, propelling himself painfully and laboriously with elbows and knees. There was no sign of the old guy ahead, but in spite of the thorns and snags tugging and ripping away, Geza tried to move as quickly as he could. He was determined to keep on following.

At last there was a sign. Or signs: a sound, a rustle, a scribble-scrabble hustle growing in intensity. Geza soon understood that whatever or whoever made the sounds was headed his way down the tunnel and could not possibly be a chimp. Ahead of him, now, he saw dimly some rough beast perhaps a dozen yards away, filling the tunnel and coming his way like a cannonball. It was a bushpig, a one- or two-hundred-pound projectile of muscle, gristle, and bristle armed with sharp tusks and running stiff-legged at top speed. Geza froze, waited until the last second, then shouted fiercely and rolled against the left side of the tunnel, stretching his arms and legs straight out and back. The bushpig rocketed past—followed by three youngsters, who were themselves followed by a rank stench—while Geza continued to lie there, stretched out and shaking.

Once he recovered his nerve and stopped shaking, Geza returned to crawling through the tunnel. He emerged at last on the lower slope of Dung Hill—scratched and bleeding, his clothes torn—to discover

Worzle sitting in a tree and gazing intently at him with what seemed to be curiosity. Maybe it was those humanoid eyes of Worzle's, eyes that gave him a thinker's face and that made him seem now like a canny planner, a trickster who took a quietly bemused pleasure in Geza's scratched-up condition and his confrontation with the racing beast in the tunnel. Had Worzle intended any of that?

. . .

Another time, Geza was following two chimps. One was Figan. The other, to the best of Geza's recollection forty years later, was Evered. It was one of the B-record follows, although *follow* may not be the best way to describe what was going on. All three of them were walking single file on a path and going in the same direction. But Geza was in the middle. Evered was in front. Figan was behind. It was not an abnormal way to travel. Human followers often got mixed that way into a group of chimps traveling in the same direction.

The trail they were moving along went out to the Kakombe waterfall in one direction and, in the other, down to the main camp. They were headed toward camp and were not far from it, maybe only fifty or a hundred feet away, when suddenly Geza heard a faint scratching or rustling in the underbrush. The chimps must have noticed the same sound at the same time. They had seen monkeys recently, not far away, and Geza thought the scratching or rustling was caused by a hidden monkey. Perhaps a red colobus, the Gombe chimpanzees' favorite source of meat. Immediately, the two chimps acted as if they too had concluded it was a monkey. They became excited, and the hair on their bodies rose as the chemistry of excitement coursed through their veins. They also became quiet. They left the trail and fanned out, Figan moving to the right, Evered to the left. Their movements slowed down, and they focused intensely on that one spot. Geza knew what was happening. They were stalking. And he stood there, staying back on the trail and doing his part as the objective scientific observer waiting to see what would happen next.

Both chimps stood upright and crept forward slowly, slowly, slowly toward the part of the thicket where the noise had come from. When they got within two or three feet of it, they stopped. Then Figan turned his head and gave Geza *the look*. Chimps don't necessarily look at you the same way in different situations. If you don't know chimps, you can't tell the difference, but when you know chimps, and especially when you know individuals, you can reasonably interpret those looks in precisely the way you would interpret the subtle signals involved in any person's expressive look. The look, Geza thought, registered a severe complaint. It seemed clear to him that if Figan could speak, he would have said something like: "You shit!"

By standing there and doing nothing, he was ruining everything. So at that moment, and in response to the look but without really thinking about it, he stepped off the trail and moved forward, preparing to back up the other two as if they were a hunting party of three. It was the natural thing to do, and both Evered and Figan, now apparently satisfied that the two-legged ape was contributing his share, returned to the business at hand: creeping closer and closer to that mysterious, hidden source of noise, stalking whatever or whoever lurked in the bush.

What happened next terrified Geza, because it was not a monkey the three of them had been approaching so stealthily. In an instant, the leaves of the bush lifted and the bush opened up entirely, as if someone had violently yanked on a cord of venetian blinds, and a Nile monitor lizard sprang out and raced straight toward him. Not them. Him. Nile monitors at Gombe can be six feet long. This one was not that big, but he was big enough, and he was upright and running on his hind legs. The reptile rushed at Geza with tongue extended—it seemed a foot long—while hissing loudly. Geza almost peed in his pants, he was so shocked and scared, and an image was permanently seared into his brain: giant upright lizard with tongue out rushing between two chimps, the two chimps staring right at Geza, their mouths hanging open with shock and amazement.

He ran, just ran. Got behind a tree while the lizard raced right by and disappeared with a slashing sound through a dry thicket of leaves. And then, after the three of them—Geza, Figan, and Evered—regained their composure, they returned to the path and resumed walking together in the direction of camp.

3.

Ruth's passion for geology as a discipline was one logical response to being more comfortable outdoors than in, more content with nature than with people. Her interest in nature, in being outdoors, had been clear in Washington as well, which was true despite her coming from a family where nature was not part of the equation. No camping. No hiking. No picnics. None of that. So perhaps that powerful interest, that attachment to nature, was simply part of her given personality. No one had much time off at

Gombe. What free time Ruth and Geza had together, they usually spent going out for walks, exploring places they hadn't been to before. Ruth was always making comments about the rocks they were walking over or the formations they looked at. Geza understood her interest in geology. He had acquired a geological perspective because of his father, but he did not have Ruth's passion for it.

Both of them found it exciting that the Gombe research camp was perched precariously on the rift escarpment, teetering on the moving edge in a geologically active part of the world. Because of the looming escarpment, that geological fact, they woke up every morning to an anticipatory dawn that would hesitate and hesitate before gloriously shifting into day once the sun reached the requisite height and angle and began to pour molten gold all around them. Even the quality of the morning light reminded Ruth of the massive forces below, and Geza remembered quite clearly her joy at experiencing her first earthquake. Most people have some anxiety when they experience an earthquake for the first time. It was the opposite with her. The quake happened in the middle of the night. Some people didn't even wake up, but Ruth did, and she was excited by the experience.

And yet Africa, or Gombe, brought her something else, something different and deeper. Gombe changed her, and it may have done so in part because she found a place and circumstance where her faults became virtues. At Gombe, Ruth's deep reserve and wrenching shyness, elements so often regarded as pathologies to fix or overcome, turned out to be strengths. Her outer reserve was sustained by an inner intensity, and while she seldom initiated conversation, that very lack of what might seem like the *Homo sapiens* babbling gene suited the chimpanzees just fine. In addition, her intense shyness around people had already made her, as a matter of unconscious compensation, a superb observer. That was Geza's opinion, anyway, and I can readily accept the theory that Ruth had already, before she came to Gombe, developed the capacity to look rather than listen, to read the half-hidden language of gestures and postures and expressions, the encrypted code of the body with all its complex layers of approach, withdrawal, feint, and deceit. That, too, was an aspect of the social world of chimpanzees. The chimps were excellent communicators among themselves, even though none of their communication was cast in the form of spoken symbolic language.

As the months passed, Ruth found herself swept away by the drama and beauty of where she was. She began to fall in love with the wilderness and the wild animals in it, especially the chimps. That was an attitude or feeling she shared with Carole—and, over time, increasingly with Geza.

Ruth had been Geza's friend and lover before coming to Gombe, although it was never clear to either of them where that relationship was headed. But being at Gombe opened something in Ruth that had previously been closed, and as she came to feel more comfortable with the conditions, and to feel increasingly attached to the place and its elusive, nonhuman denizens, she also became more sensitized to her one important human relationship. That was another surprise. Geza, in his own version of shyness, was too verbally circumspect to use words like *love* readily, but it is clear enough that he, too, was shifting into a more intense phase of personal attachment.

Such interior tectonics took place in a situational context: of isolation and intensity. Both Geza and Ruth considered the isolation of Gombe a positive thing. They never felt a need to do the usual American things, like go see a movie or eat hamburgers in a restaurant.

The intensity was part of the culture. A naive visitor might have called it obsession—mostly with the animals and what they were doing. It was like living inside a soap opera, where it seemed as if everything people talked about related to the characters in the drama. Indeed, they worked so hard for such long hours that even had they been inclined to concern themselves with other matters, there was little time for doing so. They put so much effort into their work not because Jane instructed them to do so. Jane rarely instructed anyone to do anything, which was probably part of the reason why people worked so hard. They would have rebelled against too much control and interference, and so she got more work out of people by leaving them alone. They were all doing as much as they could, and the hours were staggering. Up at five o'clock, ready to leave camp by six, following the chimps for as long and as far as possible, typing up all the notes for whatever they had done that day, and often working until eleven o'clock at night within the flickering halo of a kerosene lantern. What they did was beyond anything you would expect from a normal work crew being paid to do a job, and of course, no one was paid much anyway. The monetary compensation they got was barely enough to put something aside, each month, for a decent holiday at the end of the year plus, maybe, a few packs of cigarettes now and then.

. . .

For the first two or three months, Ruth had been afraid of the big male chimps, which was entirely reasonable. Even the other chimpanzees—

the smaller males, the females, the youngsters of both sexes—were afraid and occasionally terrified of the big males. But as she gradually overcame her fear of them, Ruth became fascinated by and attached to those same big males above all.

There was Faben. When he emerged from the forest and showed up in camp in early July, after a long and mysterious absence, it seemed as if several of the other chimps hadn't seen him for a long time either. They greeted him so enthusiastically, running up and embracing him wildly, that the sight just made Ruth choke up with feeling. She felt like crying. Like Carole, Ruth thought Faben was a really attractive ape. "Outwardly he is very handsome," she wrote to her parents, "but besides that, the expression on his face and the look in his eye just fills me with awe. I can't quite say why; it just does."

Then there was Humphrey, who happened to be, so Ruth wrote home in June, "one of the biggest and most impressive of chimps," someone who "looks like something out of the movies!" One time while she was taking notes at the provisioning site on a nonbanana day, Humphrey seemed to become incensed by the absence of bananas, and he suddenly rose up and charged her in a great fury. It was frightening. His hair bristled fierce and shiny, and he stood on his hind legs and ran right at her—only to stop abruptly four feet in front of her. He stood still and waved his arms at her. In truth, Ruth soon decided, Humphrey was not really someone to be feared. And in fact most of the big males were harmless—except, that is, for those two or three irascible old weight lifters who sometimes grabbed, slapped, kicked, or occasionally bit her. But even the grabbing, slapping, kicking, and occasional biting were pro forma: glancing blows and gestures done restrainedly, as if she were not particularly relevant or even really there. For some strange reason, the males' attacks on people were never perpetrated with anything near the intensity of their attacks on other chimps. After Humphrey stopped abruptly in front of Ruth that time, he stood there and waved his arms at her, then turned and went after a female chimp bystander. Grabbing her by the hair on her back, he dragged her down a hill and, as Ruth wrote, "continuously stamped on her and practically beat her brains out." Why was Humphrey so restrained with her and not with females of his own kind? It was just one of those strange things.

"Now that it's over," Ruth wrote in a letter of July 27, after she had been living at Gombe for more than two months, "I can admit that I really was . . . afraid at the beginning of certain male chimps." As a matter of fact, there was a time when she thought she would never overcome those fears. But then suddenly, in a few weeks' time, she came to know them more fully and regarded those same males as "among my favorites."

The worst became the best, in other words, and the very worst and then very best of the lot turned out to be Charlie and his older brother, Hugh. They both had magnificent physiques and impressive charging displays that they sometimes performed in parallel. Unlike most of the other males, moreover, Charlie and Hugh regularly tried their best to unnerve people during their displays. Ruth wrote,

> I was scared to death of these at first and couldn't even tell them apart. Actually they look as different as day and night, but I must have had some kind of mental block. One of the first researchers used to play with Charlie, and people think that is why he acts the way he does today. Nobody, however, played with Hugh. Neither one is at all vicious, they just like to grab ahold of you or else try to push you down the slope! I've never had a run in with Charlie, but have several times with Hugh, but he has never hurt me. I'm at the point now where I almost wish they would come at me because I like them so much!

None of the other researchers liked those two very much. Not Geza. Not Carole. Not Patrick or Alice or anyone else. But Ruth did, and her experience with them and some of the other big males was, she continued in the same July 27 letter, "getting deeper and deeper," to the point that she was "becoming so involved with these animals that I never want to leave them." Charlie and Hugh never let a human being follow them out of camp. No one, that is, except for Ruth. They let her follow. In fact, Ruth had begun to believe that Hugh actually liked her in much the same way that she liked him. One time when she was following the two of them into a deep valley, Hugh, during a momentary pause and rest, came up and sat down right next to her. The act seemed deliberate, and it appeared to communicate a casual, offhanded trust. Another time, he placed himself directly in front of her, seemingly indicating that he expected a friendly grooming. Presenting himself like that "almost broke my heart," Ruth wrote, since it was against the rules for researchers to have any physical contact with the chimps, and so she felt obliged to get up and move away.

By the time September came around, Ruth had begun to think that Hugh was actually her favorite chimpanzee of all. One time near the end of the month, when she was following him high into the hills, she thought she had "never seen him more magnificent." He was fierce and proud, swaggering, full of himself and "positively full of the devil—you could see it in his eyes and in all his movements." He stood up and displayed on and on, passing through the vines and tangled vegetation as if they were mere shadows instead of actual obstacles. "I don't know how he could have done it," Ruth wrote, "as the vegetation was very thick and at least waist high. But he displayed through it with no effort whatsoever. He then spotted me and came quite aggressively toward me. He didn't push me over today (!) but sat a few feet away staring at me. How fantastic he was! After several minutes of this he came over and presented [for grooming]. I can never express how awed and thrilled I am that he chooses me to do this to. And the utter agony of not being able to reciprocate."

Near the end of October, Ruth followed Hugh into a thick section of forest when a fierce storm began to move in and everything became strangely dark. That part of the forest was typically dark at that time of day, late afternoon, but now it became preternaturally dark, and Hugh, perhaps stimulated by the atmospheric shifts and the rising wind and dimming light preceding a storm, began to show off his magnificent physique and power with display after display. He would run upright, the hair on his legs and arms and back and neck bristling, and he began beating or slapping on his chest like a gorilla, throwing big rocks, then drumming resonantly on the high, thin root-buttresses of certain trees. The drumming was loud and fast and terrifically exciting, and the trees looked ghostly and glowing. When the storm finally arrived and the rain began splattering, splashing, and then pouring down, Ruth and Hugh huddled near each other, and together they got cold and soaking wet. "Having spent so much time alone with Hugh in the woods," Ruth concluded, "I am getting to know him in a much deeper way than I know the other chimps." He, along with Mike—the community's alpha—had become "two of my most favorite animals, in fact, two of my most favorite beings in the entire world!"

4.

The absence of Carole that fall meant that Geza and Ruth did all the B-record follows. The chimpanzees often moved in groups, and the two people would sometimes pick as their target individuals two chimps who were part of the same group, which gave them the luxury of following together. One evening in early November, they jointly followed some chimps into an area just north of a grove of *myombo* trees on the slope above camp. There the apes constructed their sleeping nests and retired for the night. Geza and Ruth returned to camp, rose early the next morning, and, after a gulped-down breakfast, stepped outside into the faint morning light. Soon they were standing silently beneath the still sleeping animals in their nests.

At around seven o'clock, first one then another faint silhouette emerged from the high nests and passed quietly into the overhead chiaroscuro. Against a speckling of light through leaves and branches, dark shapes moved here and there, climbing onto and riding down an elevator of supple saplings, swinging down on branches and vines, or clambering hand over foot down a tree trunk until, one by one, they reached the ground and turned from shadows into apes. There were seven altogether: six males (Mike, Hugo, Goliath, Godi, Dé, and Goblin) and one young female (Melissa). Yawning, blinking, stretching, sitting quietly in the predawn mist, still shaking off the night's sleep, they lingered for some time before starting the business of the day.

With Mike and Hugo taking the lead, the group turned into a parade of swaying shapes moving east along the steep side of Peak Ridge in the direction of the escarpment top, passing through a silvery haze and then, suddenly, facing the first blinding burst of sun. Mike and Hugo turned down in the direction of Kakombe Stream, with the five other chimps and two humans following behind. They then proceeded east in the direction of Kakombe falls.

Progress was a meandering enterprise, with various individuals stopping to rest or stepping faster, slipping behind a curtain and disappearing before reappearing somewhere else a few minutes later. The group passed into and through open woodland that gradually changed into mature forest: shadows under a green canopy, old trees wrapped in strangler fig, vines and lianas looped wildly about, and a ghostly fluttering scatter of large white moths. Time passed. The apes coalesced and paused to rest and scratch themselves, to groom one another, to exchange casual greeting grunts, to gaze abstractedly into the buzzing morning.

A few quiet minutes passed before Ruth and Geza heard, from some obscure place in the forest ahead, an echoing exchange of pant-hoots. These were familiar, friendly voices that brought everyone of the resting group into motion once more; and soon the moving group was joined by four more males (Faben, Evered, Satan, Sniff) and a young female (Winkle), so now Ruth and Geza were members of an ape parade that included a dozen chimps. Then Humphrey appeared and joined the moving group, making it a baker's dozen, although he soon disappeared along with several others into the forest ahead. Ruth hurried on, hoping to keep Humphrey in sight. Geza stayed back to keep track of Melissa and Goblin at the rear.

So far it had been an ordinary day's ramble.

At 8:38 in the morning, however, just before Geza and Ruth reached the foot of Dung Hill, they were alarmed by a dramatic explosion of sounds ahead: a shattering of hoots and screams, *waa*-barks and wailing *wraaahs* coming from not far away. The auditory chaos seemed to express, as Geza would later write, "an oddly hysterical mood"—particularly the *wraaah* calls, which are capable of carrying "for miles along the acoustic funnel formed by steep valley slopes." Radically unlikely any other vocalization, the *wraaahs* were "emitted in series as long, drawn out, high pitched, clear toned, plaintive wails that instantly alert everyone, observers included, to danger." Faben and Hugo, their hair suddenly spiked with excitement, bounded on ahead, while Geza raced to catch up with Ruth. They exchanged glances, and then, wordlessly, they both turned and ran in the direction of the alarming din.

They arrived at a copse of tall fig trees where several dark apes, hair bristled and shiny, careened in one direction and another, racing full tilt on all fours or completely upright and using their arms to pivot, to tear, to throw. They were screaming and hooting, *waa*-barking and *wraaahing* at full volume, which was loud enough that the human observers' ears were ringing, their legs shaking with the intensity of it all, their hands trembling. They took in the scene, recognized three additional newcomers to the group: Hugh, Charlie, and Willy-Wally. Then they saw the dark heap. After fumbling with the tape recorder switch, Ruth spoke into the microphone, stating for the B-record: "Hysterical group of animals . . . frenzy of screaming, wraaahing, and displaying . . . dead body of Rix lying face down in the gully."

Melissa and Winkle, the only females present and both small enough to be vulnerable to the males' violent displays, scuttled anxiously about from place to place, seeking temporary refuge behind tree trunks. Young Goblin circled about helplessly, also apparently trying to avoid the big males who still hurtled about. Godi and Charlie perched in some low vegetation, staring at the dead body and emitting continuous *wraaahs*. But the other males were just raging.

Mike dragged and whipped tree branches, then shook a hissing cutlery of palm fronds. Hugo hurled large rocks in an underhanded toss as he ran. Humphrey hunched over and pounded the earth with his feet before catapulting himself forward with his great arms and slapping his big hands onto the ground. But these displays, as violent as they may have seemed, didn't express aggression, Geza thought; rather they were "designed to provoke movement in Rix." To revive him, in other words. "His stillness clearly bothers the others, and rushing by near the corpse may release tension while also testing whether Rix is really dead." And even as the males careened around the clearing in their explosive displays, passing within inches of the two humans trying to sputter accurate notes into their tape recorders, none of the chimps ever directed a slap or blow or any other aggressive act or gesture at them. It was extraordinary for the two humans to witness this wild, wild, wild behavior, while the chimps, seemingly lost in the intensity of the moment, hardly appeared to recognize their existence. And at the center of all this raging action lay Rix, his body heaped in a shallow gully and turned on its right side: an inert cluster of hair and flesh with unmoving hands, legs folded into the chest, a frozen face, eyes gazing blankly into an alternate universe.

Several rocks protruded from the forest floor where he lay, one of them big and sharp-angled. A great fig tree towered almost directly over the rock and the body, and the tree had no significant branches lower than thirty feet off the ground, which meant that Rix—presuming he had fallen from the tree—had nothing to grab on to on the way down. Perhaps he had fallen and struck his head on the rock. His neck was severely twisted, suggesting that it could have been broken in the process.

It might have been an accident. For a species so brilliantly competent in the trees, an accidental fall would be unusual but not unheard of. Or perhaps Rix had gotten into a fight. Perhaps one or two other chimps had knocked or pushed him during a struggle high in the tree, and he fell. Or he might have been surprised, unnerved, and then unbalanced by an aggressive display or a charge made by another male.

By 8:53 A.M., about a quarter of an hour after Rix had seemingly fallen or been pushed out of the tree, the raging chaos began to settle down. Most of the chimps were now sitting, grouped in small clusters, although still close to the body. Some had begun to groom one another quietly. Some were gazing intently at Rix's face. Godi had regularly been seen in the company of Rix, as had, slightly less often, Dé and Willy-Wally. Godi, in fact, kept making *wraaah* calls, with Sniff and Satan occasionally joining in to produce a chorus of *wraaahs*. Perhaps Rix's death, that sudden existential suck of being into nothingness, struck Godi as an unacceptable transformation. Godi was perched on a branch slightly above the corpse, balancing himself with one hand while staring at the body. At times, he reached an arm out in the direction of the body, palm turned up in a begging gesture. He looked upset, and he repeated again and again those anguished calls. Godi's *wraaahs* seemed to Geza mournful. That was the word that came to mind. The whole thing gave him chills.

By 9.37 A.M., the young males Godi, Dé, and Willy-Wally had calmed down, and it now seemed to Geza that those three were guarding the body. But Sniff moved in close, brought his face right up to Rix's face and poked at an unmoving left arm. Then Humphrey approached the crumpled heap in the shallow ravine and inspected it. Gradually, however, individuals began drifting off, wandering away. Some left and then returned, as if reluctant to turn away for good. Godi was the last to go, finally, at 12:05 P.M., departing with a slow shuffle and slipping into the forest south of Kakombe Stream.

5.

The winter rains came and made late November the season of the strange blossoming. At dusk, certain dark and towering castles in the forest began to pour forth thousands of winged insects: working members of a species of termite (*Macrotermes bellicosus*) living in dark, multichambered nests deep underground that were protected aboveground by those hard, earthen castles. The castles were defensive fortresses erected in a supremely hostile environment. All sorts of eager enemies were waiting outside, ready to snatch the flying termites right out of the air and eat them. Monkeys did that. Birds. People. Chimpanzees.

Those same besieged colonies included another caste of individuals who were much fiercer and more difficult to find and catch than the flying workers. They also happened to be bigger than the workers and to taste good. These were members of the *Macrotermes bellicosus* soldier caste: aggressive biters with powerful mandibles who lurked forever in the castle darkness protecting the queen, never rising voluntarily into the light.

They were vicious but not especially clever and could be drawn out by a craft and trickery practiced primarily by the chimpanzees. If a chimp selected just the right blade of grass or palm frond, or an especially long and supple twig, and fashioned it just so until it became a long and flexible probe; and if the same individual applied a certain virtuoso technique to slip and snake that probe deeply into the castle tunnels and provoke the soldiers into clamping on with their mandibles; then that very ape could carefully draw out the tool to behold a mass of clinging termites vulnerable to a quick swipe with fingers into mouth.

Not so easy, this fishing for termites, but any good chimpanzee at Gombe learned to do it as a bumbling youngster among skilled elders. The youngsters watched and imitated and were guided and sometimes actively taught to do that.

Back in 1960, Jane had famously discovered that the Gombe chimpanzees fished for termites, and now, eight years later, Geza and Ruth spent a good deal of time observing the same behavior. The chimps made it look easy, but it took significant skill to navigate those narrow, twisting tunnels with a flimsy blade of grass or palm or a twig. Geza knew how hard it was because he tried and practiced it, using Leakey as his skilled mentor. Leakey showed him how.

"I have been termiting with Leakey many times," Geza wrote to his friend George Rabchevsky that December, "and am sometimes more successful than he. But usually he is better at it, and I have been repeatedly stumped by the way he can go to a completely earthen mound and immediately find a place where a termite hole is near the surface." It was true, Geza continued, that "sometimes I get more soldier termites than he does, and then I always give them to him." Other chimps, when they were outdone like that, would "get mad and chase you away from the good hole," throwing tantrums or rocks, "but never old Leakey. He just fishes at his hole and also eats the termites which I hold out for him, thus getting twice as much."

. . .

Geza's letters to George were almost always breezy and cheerful, and in that way they seldom represented the author's full range of feelings and concerns. He was, for instance, troubled during this same time by the letters he was receiving from his draft board in New York. The board had refused to give him permission to finish his research and now were threatening to reclassify him as delinquent and, presumably, to send him off to Vietnam if he didn't return to the States for a preliminary physical examination. Meanwhile, Christmas was fast approaching, and that fact, combined with an oppressively overcast sky and wet weather, brought a nostalgia-tinged sense of deprivation. "The research station here is good for many things, but I can think of better places to spend this time of year," Geza admitted in a letter written that December to his graduate advisor back at Penn State, adding that, "aside from the problem and expense of getting things in Kigoma, even food, the rainy season has come upon us with a vengeance and with all its physical discomforts."

Nic and Margaret Pickford had returned by then, having driven the VW bus back from Ngorongoro Crater in the third week of November and bringing with them a long-term volunteer from America named Cathy Clark. She had spent almost a month at Ngorongoro Crater getting to know Jane and Hugo, which was long enough for them to develop a strongly positive opinion of her. "She's one of the nicest girls we've had," Jane wrote to the family in England. "We both like her most frightfully and, unlike most of these Americans, she's prepared to rough it, and does not expect caviar twice a week—nor does she take vitamin pills." Patrick McGinnis was also back. He'd had a satisfactory visit with his family in California and a wonderful stopover in Cambridge, England, where he was to begin working toward his PhD in January. The small plane that brought him from Nairobi to Kigoma, meanwhile, picked up Alice Sorem for the return trip. Having decided that she was finished for good with Gombe, chimps, and research, Alice married the pilot who flew her out.

Nic was pleased to be back, though, and, while chasing snakes in his spare time, he captured a strange serpent none of the experts in Nairobi could identify. Maybe it was a new species. Maybe even a new genus. To the discoverer went the naming rights, and Nic began to imagine the scientific name he would choose. The new snake species would be a *Pickfordi*. The problem was that no one could tell whether the particular snake he had found was an anomalous freak, a strange hybrid, or truly the first discovered member of a brand-new species or genus. He would have to provide more *Pickfordi* specimens for anyone to tell, and so far he was now finding only the already-known kinds of snake. As for Christmas, Nic, like Geza, believed it was likely to be a dreary one.

. . .

Maybe everyone felt that way, but then Hugo made a surprise visit a day or two before the holiday, flying in from Nairobi to deliver a turkey, a Christmas pudding pickled in brandy, and a cake. He also brought the ingredients of a second feast for the staff, who had the day off and would be relaxing with their families in the staff village.

Early preparations for the researchers' dinner took place within the baboon-proof Cage at the beach. Margaret thought the cake Hugo brought looked tiny at first, but she soon transformed it into a model of Gombe complete with beach, mountains, tiny bananas (made from dried pears), and little rubber chimps (brought in from Nairobi by Hugo). People wrapped candies, snipped out paper decorations, twisted

Nic, Cathy, Margaret, and Ruth present a Christmas gift to Iddi Matata and the fishermen.

wire into candle holders, and so on. No one was supposed to cut down trees in Gombe, but Geza studiously uprooted a nonnative lemon tree that had been doomed anyway. It became the Christmas tree.

It rained for much of Christmas day, and everyone slept in. No research work required. People gave each other simple presents and partook of a cold lunch with baked ham, which was a rare treat. Nic had been authorized to dredge the bottom of the budget for something extra, so during his supply trip to Kigoma he had purchased some sort of presumably consumable treat, which he, Cathy, Margaret, and Ruth presented on Christmas afternoon to the fishermen through their spokesman Iddi Matata. Geza was there to take the photograph. Then everyone gathered at the Cage for Christmas dinner. The turkey was enormous and stuffed with an excellent dressing concocted by Geza and Ruth. Geza missed cranberries, but the pudding set aflame with brandy made up for that lack, while the cake decorated to look like Gombe with bananas and chimps on top was fun and satisfying. As was the wine. There was an abundance of wine, in fact, because Nic had found a lot of cheap Californian wine—red Gallo in big green jugs—at Ramji Dharsi's store.

Carole had been back since December 17, and although she was missing her friends in Nairobi, she was also eager to start work on the exciting new task of habituating chimps at Nyasanga. She was glad to be back. Geza made her feel gladder, since he admitted that she had been right about the chimps. Carole would remember for the rest of her life the moment he took her aside and said, privately, "Carole, I have to let you know how much I can see your point of view now. I really find that there's something between me and Leakey. It makes me know what you meant about loving the chimps." It felt good to hear Geza admit that.

After dinner, Carole recited from memory Dylan Thomas's *A Child's Christmas in Wales*. She had a small paperback copy of the story, which she had spent several hours memorizing in preparation for her recitation, and, she thought, people were really entertained. Even Margaret, whom Carole had previously considered to be among the camp's more stalwart conservatives, was enchanted by Carole's skill. Carole had been complaining about the work of typing up notes, and after the recitation Margaret said, quite generously, "I'll type any of Carole's notes."

Just about everything felt good that Christmas, and the Gallo wine probably helped. After dark they began to play tape-recorded music down at the beach, and Carole and Nic began dancing. That felt good, too, and soon enough they had spun away from the lantern's shivering embrace and, still dancing, slipped into a winding tunnel of darkness. Other than that, nothing happened. Nothing, that is, except perhaps for certain indefinite sensations and feelings that may have been experienced during the brief passage of time before some of the others ran out and thoughtfully tugged the two of them back into the light.

6.

A few days after Christmas, Carole moved down to her new place at Nyasanga. Nyasanga was the name of the stream that gathered in its watershed near the top of the escarpment. It dawdled and puddled, grew and flowed, shivered and dropped its way down to the lake, in the process and over many millions of years carving out the Nyasanga Valley. Carole began to think about the names Nyasanga, Kahama, Mkenke, and Kakombe—four streams and four valleys, arrayed like a series of complicated passageways running east to west, sectioning off the Gombe terrain from south to north—and the names became like street names to her. Nyasanga. Kahama. Mkenke. Kakombe. They were streets to count off whenever she felt too much by herself, which happened often during her first several weeks at Nyasanga.

She had never been that far south before, and now she had to spend her days in that terra incognita when everyone she knew and felt any sympathy for was up in the main camp at Kakombe. She always had that wonderful feeling based on her love of Gombe's beauty, but down at Nyasanga she sometimes felt merely dull and frightened, partly because the valleys to the south were narrow. There was no peak like the Peak behind the camp at Kakombe, that high lookout where a person could raise binoculars to her eyes and peer into the forests all around, could scan across a large area looking for the characteristic boil and swirl that chimps and monkeys make in the trees. And because the southern valleys were narrow, a person couldn't hear things over long distances. Nyasanga and the other southern valleys were nothing like Kakombe Valley, which was a wide open bowl in which a person could hear chimps calling from far across the valley. Nyasanga was harder territory. The whole escarpment there was squeezed down closer to the lake.

One thing made Carole's loneliness at Nyasanga more tolerable: the weekly appearance of Nic Pickford. Nic was responsible for the weekly shopping trip to Kigoma, after all, and an important part of that responsibility was bringing food, supplies, and mail to Carole at Nyasanga. So Nic would stop by, stepping out of the boat and into the water, hauling the boat crunchingly onto the shore, pulling out boxes of supplies and bundles of mail. Carole soon found that on those days she would at a certain time begin listening for the whine and chop of the boat and waiting for that gentle smile accentuated by the ring of neatly cropped facial hair, the reassuring voice, the uncomplicated practicality, the inherent sweetness and need to be liked. Nic was the only one she saw from camp on a regular basis, and I will imagine it did not take an excessive amount of reciprocal engagement—a casual gaze that lingered a second or two longer than expected or a glancing brush of arm against arm that felt cool but turned warm—before they became lovers.

She was lonely, yes, but she had not expected to take up with him. He was married, and he seemed very different from her. Different history, background, education, attitude, interests. And yet he was handsome and kind, and he had to stop by every week. How else was she going to get food? It was part of his job. And he was interesting. He was a good snake man, skilled with machines, handy with a camera. He was basically, Carole thought, *a good bloke*. Carole had learned a lot about men and males from watching the chimps and from being in the forest, and she possessed more self-confidence now than she used to. She still felt attached to the

Nic was handy with a camera.

Gangster Boyfriend in Nairobi, certainly, but she had also developed this foolish idea that you could love more than one person at a time. In spite of what some people considered to be her casual approach to life, though, Carole was never happy with the situation. Nic was married to Margaret, and Carole recognized that in loving Nic she was committing adultery. She began to think of herself as an Adulteress with the capital "A." And yet she was swept away by something greater than any of the more sober ideas she maintained about what a person ought to do or be. She was overwhelmed by a need for physical contact, and I believe it was not too many afternoons of being with Nic in that way, of fulfilling that need, before the experience was transformed into a radiant pleasure that was even, she must have seen, threatening to turn into love.

. . .

At Nyasanga she lived in a prefab aluminum hut with a concrete floor and, on the roof, an insulating grass thatch held in place by chicken wire. It was much like her old hut at the research station at Kakombe, the rondavel not far from the dinner pavilion on the ridge. And as in her old place, in this one she would occasionally wake up in the night to the sounds of rustling in the thatch, which were the acoustical wakes of slithering snakes and rushing rats, the former in lethal pursuit of the latter. It was odd to consider, but that frantic froufrou of snakes chasing rats was familiar enough that it made her smile, close her eyes again, and feel better about being down at Nyasanga. She lived by herself, of course, just as she had lived by herself back at the research camp. As a matter of fact, at Nyasanga she was in some ways less by herself, since she also lived within shouting distance of a second and larger aluminum prefab, a rectangular one that had been erected by the Parks Department to house the game ranger, Ferdinand Umpono, and his wife.

Nyasanga had become the site of the official ranger station for Gombe National Park, after all, which was why Ferdinand and his wife stayed there, while Carole's job was to find and tame chimps for the tourists who would one day come to see them there. So certain people hoped. And Carole did like Ferdinand and his wife, whose actual name, unfortunately, she never learned. She only knew her by what Ferdinand called her: his *Bibi*, which was Swahili for *grandmother* or *mistress* or *wife*. Carole had to look that one up in her paperback Swahili-English dictionary, and indeed, the disconnect between Carole and her next-door neighbors was primarily a linguistic one that the dictionary was supposed to resolve but never did. Ferdinand and his wife spoke little English. Carole was locked into English, knowing only a few standard words and phrases of Swahili and having to look the rest up.

Ferdinand never presented himself as a bold or impressive man, and he was not. He was skinny, and he wore game ranger khaki shorts that showed off his knobby knees and sticklike legs. He had a plain face, Carole thought. Not handsome. His ears stuck out a little. But then he wasn't cocky either, like good-looking men so often become, and his face was direct and honest. Nothing held back or secretive. He always wore his uniform, including a hat, and he was always clean and sharply dressed. He looked just like he was supposed to look, as a Tanzania Parks Department game ranger. Carole thought he must have been a role model, since the chief warden, Steve Stephenson, chose him to be the single full-time parks representative at Gombe. For months at a time, the fishermen on the shore, and the staff and researchers working at the research camp, knew Ferdinand as their only contact with Parks Department officialdom. So he was an important person, but Carole never got to know him or his wife well. Nevertheless, he treated her kindly and thoughtfully, and she trusted him. She was glad he was there. She knew that if she failed to return in the evening, he would report her missing and then go find her.

Ferdinand's Bibi outweighed him. She was plump and tall, while he was a small man. She had a full bosom and ample hips, and she sometimes wrapped herself dramatically in a checkered yellow robe. She kept her hair pulled back but short. Her roundish face was lit by an inner beauty that Carole believed came from contentment and an absence of fear. She acted as if she could deal with anything. She would sometimes come over and peer into Carole's window or tap on her door, try to talk to her, but that never got far. Carole told me forty years later that if she were somehow able to do it over again, she would go over regularly to their house and work to learn Swahili. Ferdinand's Bibi was trying to make friends and get to know Carole, and, as Carole eventually came to recognize, Ferdinand and his Bibi didn't have any real friends or relatives in the area. They were both outsiders. They weren't Waha. They came from a different region and tribe.

But Carole was so caught up in her own existence, so snagged with incessant rehearsals of interior reflections and concerns, that she didn't have much extra energy for cross-cultural socializing. She had books to read, even though she didn't have a lot of extra time for reading either. At the end of the day, she would come back tired. She would fix dinner for herself. Read a bit. Go to sleep. In the morning, she would get up and go out again, looking for the chimps.

True, she was still excited to be doing something comparable to what Jane had done in her first pioneering year or two: getting to know a new piece of virgin forest and meeting the new chimps, discovering and becoming familiar with them. But she had thought finding the chimpanzees at Nyasanga would be easier than it was. She had believed that some of the familiar chimps—her old pals, her good old hairy friends— from the research station at Kakombe would be wandering that far south, maybe mixing socially with other chimps down there at Nyasanga, and so she imagined that those familiar chimps would in essence introduce her to the stranger chimps down south. She believed that the old chimps she knew would relax and feel comfortable when they saw her, and that their relaxation and comfort would positively influence the reactions of the stranger chimps.

Unfortunately, though, that hopeful idea was based on a misunderstanding of chimp society. Later studies would clarify that chimpanzees are xenophobic and territorial. They form territory-based communities that are, in essence, isolated little nations, and the boundaries between those nations are maintained by the adult males. Young, nubile females sometimes emigrate from their home territories and cross those territorial boundaries, thus moving between communities, but they do so only in early adolescence. No one else does. If chimps had religion, in other words, each community would worship a different god. If they had flags, each would hoist a different design. In short, the chance of Carole finding sympathetic and influential old friends from the Kakombe area among the chimpanzees living in the southern part of the park was remote. She was exploring the lands of another nation.

That was one problem with the southern chimpanzees. Another was: Where were they? For the first several weeks, Carole saw chimps only every once in a while. She kept looking in Nyasanga and the valley just to the south, Kalande, mainly because Steve Stephenson had decided that chimps were there in those two valleys. Perhaps for that reason he had chosen Nyasanga as the location of the new tourist station. He had told Carole that there were loads of chimps at Nyasanga and Kalande when he came.

The one day he came.

As Carole later deduced, a particular fruit, probably *mgwiza,* was in season at the time, and the ripe fruits drew the apes in close enough to the lake that Stephenson heard them from the beach. That was the source of his loads-of-chimps idea. So Carole kept looking again and again in the same places, and she kept finding nothing. No chimps. Oh, occasionally one or a few would be there, but rarely. In addition to the absence of chimps, meanwhile, was the presence of rain and rain and rain. The rain came down like little monkey fists, and it leapt up like startled mice. The rain persisted and became a heavy curtain. The rain turned into mist and moist and mold, and it became a silver silken presence. The rain became a madman pounding on the roof of her hut in the morning, and so she would stay in bed and listen. The rain meant that she couldn't hear much anyway, and moving about was twice as hard as normal. The rain made the vegetation swell up and become taller and thicker, and it melted the ground and made it slick and slicker. She slogged. She slid. She'd take a step and fall in the mud. It was wet. It was cold. It was not nice.

Carole had taken some of the Ujiji marijuana with her down to Nyasanga, a whole big canister full of cleaned and beautiful grass. It was good, and when she smoked it, she enjoyed the effects. She could smoke it in the evening and feel relaxed, and on those occasions when she smoked during the day, she would sometimes feel more like an ape, a human ape, and she thought she could understand more about what it meant to be a chimpanzee ape. She imagined that the marijuana drew her more deeply into the truth of things. Those were the good effects, which happened first. As time passed, though, she came to discover the bad effects. Instead of making her relaxed, it made her tense and anxious.

Then there was the isolation, which soon, and in spite of the weekly wrestling and therapy sessions with Nic, began to make her days longer and more difficult than they would otherwise have been. The isolation affected her mood. Some people would have been able to tolerate it better than she could, but she was young and had just left her Gangster Boyfriend, so she carried that loss around with her. She would go visit the main camp at Kakombe once in a while, but it wasn't easy getting there. It was about a five-mile walk along the shore to Kakombe. Geza and Ruth were there, and others—Tim and Bonnie Ransom, for example—whom she wanted to see. In fact, Carole, while she liked seeing Geza and Ruth, was mostly interested in visiting with her two good friends there, Tim and Bonnie.

Only now it was one good friend there. Bonnie had gotten sick in December and gone off to Nairobi to see a doctor. She still wasn't back in January, which meant that at first Carole and Tim had, among other mutual interests, Bonnie and her absence to talk about. Later on, when it became clear that Bonnie did not intend to come back, they had more to talk about, except that Tim was upset, which made their talks less satisfying. And then, after it became clear that Bonnie was permanently gone and intended to leave Tim for good, he was devastated. It was hard for him to concentrate on the baboons and their lives, and it was equivalently hard for Carole to have uplifting conversations with him. Carole felt she had lost one good Gombe friend, Bonnie, and was rapidly losing touch with her other good Gombe friend, Tim.

7.

Two new people showed up at the main camp that January, both from Cambridge University in England—Michael Simpson and Timothy Clutton-Brock—and so when Carole occasionally walked up to join the others at dinnertime, she also interacted with them. Michael already had his PhD, and Robert Hinde had sent him to Gombe as the long-awaited senior scientist. Being a former student of Hinde's at Cambridge meant that Michael was thoroughly educated in the European tradition of animal behavior science (ethology, as it was called), and his primary task as senior scientist may have been to introduce more quantifiable methods of data collection. More quantifiable: putting away the tape recorder and the typed narrative, picking up the pencil and ticking off boxes on a sheet of paper to create reliable columns of data ready for crunching.

Tim Clutton-Brock was planning to carry out his own study on the feeding behavior of red colobus monkeys in the ethological style, but who else at Gombe might the new senior scientist instruct or influence? Michael did not seem, as a matter of fact, especially senior to the others in age or experience, and his dissertation research had focused on Siamese fighting fish. He was also exceptionally shy—much like Ruth in that way—and to Carole he seemed to spend a lot of time down at the lake looking at fish. Then he would go visit with Tim, who was the more confident and outgoing of the two by far. They were both products of an elite university environment that prized discussion and debate, but Tim was the one who liked to talk, and so it was Tim who would work on stimulating conversation during dinner.

Tim was tall with curly blond hair and bright blue eyes. He was good-looking, polite, and easygoing. He had spent a year in Africa already, working as an assistant ranger for the Game Department of Zambia, and, as he once told me, that experience taught him that small luxuries when living in the bush are important. He had moved into Carole's old rondavel at ridge camp, and at the first opportunity he acquired a small paraffin refrigerator for the place. Then he mail-ordered a blue plastic inflatable armchair. He got music tapes: Beatles, Rolling Stones, Bob Dylan. The quality of the sound was not good, but playing tapes was the only available source of recorded music anyone had, and people listened a lot to Tim's music. And he had a small library for relaxation reading— *Lord of the Rings* and *The Games People Play,* for example.

Bob Dylan was not a singer Nic cared for. It wasn't bad music, but for Nic it meant little. As for *The Games People Play,* Nic tried it. Couldn't read it. But then Tim understood people, and he was witty, too. He knew how to say funny things, and he knew how to put the joke on himself, which meant that he was fun to have around.

That's what Nic thought. Carole wasn't so sure. Because she held definite views about things, and because Tim liked to discuss and debate— skillfully, with polish and deliberation—she would end up arguing with him. They argued over such knotty abstractions as whether you could be objective and in love simultaneously. Carole felt that when you really loved you saw more clearly, whereas he took the position that being in love made you blind, so you couldn't be objective. And Carole, finally growing tired of such pointless talk, ended it with: "Oh, well. Wait and see." She thought the argument wasn't worth the energy. She also felt confident that Tim would eventually discover what she meant about love—and loving the chimps. To be sure, Tim wasn't watching chimps, so perhaps he would never understand what was special about them. The red colobus monkeys he intended to study, he once told Carole, were nothing but a PhD dissertation moving about in the forest. I imagine he was being rhetorically provocative in saying that, but Carole was still put off by the comment. In short, Carole considered herself a flop at conversation with Tim Clutton-Brock, but he still liked to see the words flow, and he would prod people to join in.

He tried to get Ruth talking as well, but she was so averse to phatic exchange and to argument for its own sake that he had to nudge her. *Provoke* her a little, as Carole thought of it. He would try to get her involved in conversation, but she wouldn't respond all that much.

. . .

The winter rains kept coming, and the combination of cold and damp during January stoked respiratory illnesses among the chimps. Flo and her dependent offspring—Fifi, Flint, and little Flame—all came down with a flu or colds severe enough that they simply disappeared for several days. Flo reappeared around the middle of February but without the older daughter, Fifi, or the infant, Flame. Perhaps, everyone hoped, Fifi was taking care of little Flame somewhere out there in the forest. After a few more days of everyone in camp waiting and hoping, however, Fifi showed up without Flame, and the logical conclusion was that little Flame was gone for good.

For people, the weather also made walking in the forest more dangerous. The sighs and splatters of falling water could deafen a person to other sounds. The swollen vegetation could blind a person to other sights. The rain became a gray mask, and the masking effect may have been why Ruth was slow to discover, on February 9, a hidden menace.

She had followed a large group into the forest that day. Up the steep slope to Sleeping Buffalo they went, single file, Ruth among them and climbing on all fours, until they entered a dark stand of mature forest. She had been near the back of the line, and so she passed into that twilit world moving in front of only one or two stragglers. It was already midday, siesta time, and soon the chimps began to relax, spreading themselves out on the ground or in convenient spots in the trees. While the chimps settled in for a doze, Ruth found a good resting place, close enough to the apes to keep an eye on them but still tucked within her own tight little vegetative niche.

An hour passed by quietly.

Then, abruptly, one of the chimps began moving with purpose. Ruth heard a swishing of leaves, and then she saw, in the midst of a leafy clump overhead, an adult male standing upright and staring intently at a spot in the thick understory perhaps ten yards from where she had settled. The male began to scream and shake a leafy branch back and forth. Soon the others had joined in with their own screams before hurtling themselves up into the safety of branches overhead. She saw their silhouettes rising away from the ground and into the dark arms of the forest, and she listened to a growing chorus of alarm that radiated in a swirl of screams, howls, and whimpers.

She looked for a tree to climb, but there was nothing substantial enough close by. She listened to a steady crackling as some large presence passed through the brush, and she saw leaves waving and the movement of shadow: a forest creature of some sort, surprisingly big and passing smoothly through stubborn intricacies. The creature seemed then to pause. To stand still. To paw the ground. To switch a tail. And, with a massive head lifted just so, to sniff the air patiently. Ruth stood still, hoping her stillness might make her invisible, while the chimps above continued to scream, the big males shaking branches and bristling.

Then the shadow resolved itself into a huge buffalo with an immense curl of horn and glistening black eyes, and those eyes now seemed to look right at her. The creature lowered his or her head, aimed, and plowed directly into Ruth's thicket. She crouched, coiled, and at the last moment leapt away to one side. A horn tip missed ripping into her shoulder by a few inches. As she recovered, the buffalo turned about and paused to aim again. Reaching up with both hands to grip a single thick liana drooping overhead, Ruth heaved herself off the ground while kicking her legs up until she had wrapped them around the liana and was dangling horizontally. The buffalo charged again, this time lifting head and horns and scoring a shallow cut across her thigh before turning around and aiming for another charge—again and again charging like that until finally, after what seemed to Ruth like fifteen or twenty terrifying minutes, tiring of the game.

. . .

Tim Clutton-Brock, meanwhile, was finding it surprisingly difficult to watch the monkeys he intended to study. Gombe looked to be little more than a tangle of thorny thickets with vines wrapped around trees, and the red colobus monkeys, unlike the chimps and baboons, spent all their time gathered in large troops high in the forest canopy. He tried walking the ridges between valleys, which put him up higher and made the movement of monkeys easier to spot, but he would still see them for only a short while before they disappeared into a cloud of trembling leaves. They had black-and-white faces, chestnut-red hair on the tops of their heads, red on their backs. It could be exciting when, in lucky moments, a person managed to watch an entire troop of around eighty red colobus monkeys passing through the canopy, setting off waves of swirling vegetation as they moved. But even that kind of dramatic show was useless in a scientific sense. His goal was to study the feeding behavior of those monkeys, and how could someone pathetically stuck on the ground learn anything significant about creatures passing gracefully through the treetops?

His data-collection method would help. Once he got started, Tim planned to carry around a pencil and a check sheet organized smartly with all the right columns and categories. He would carry an umbrella to keep the paper dry, and he would be wired up with a device in the ear that ticked away the minutes. He intended to follow a troop, and every quarter hour he would mark down as precisely as possible what each visible individual was eating, thus taking random samples every fifteen minutes for the full day, dawn to dark. Obviously, he would never be able to record what the entire troop was eating at any single moment, but the random sampling—of only the visible ones—should, after statistical massaging, provide reliable results.

Tim also decided to conduct his study away from Kakombe Valley, where the main camp was located and where the chimp and baboon projects were focused, and work instead in Kahama Valley. Kahama was two valleys to the south of Kakombe, one valley to the north of Nyasanga, where Carole lived. That put him in a fresh environment where the colobus monkeys were undisturbed by the other people and their various activities.

Kahama Valley was a difficult place to navigate, though, and at the end of one of his early forays there, Tim tried returning to the main camp by following Kahama Stream right down to the lake. The stream was the most direct way to the lake, he reasoned, and the shoreline was certainly the clearest way to head north and get back to camp. So he began following the stream down, wading in the water at times, periodically coming to areas where the water paused at a small lip of harder rock, pooling and then dropping down to a lower level. He scrambled over the lips the stream had created, and as he went down they got bigger. Eventually, he came to a waterfall of sorts, although it was not clear how big the drop was. There was a good deal of obscuring vegetation including some vines sturdy enough that Tim managed to climb down ten feet until he found himself stuck on a ledge hanging over what seemed like a thirty-foot drop. He had also disturbed a bees' nest while climbing down onto the ledge, and now the bees were starting to swarm, so: *Which way to go?* Should he jump down that thirty-foot drop, hope to grab on to some vines or tree branches to break his fall, or should he turn around, go back to the swarming bees, and try to climb back up? He took a chance with the bees, got stung a half dozen times, clambered back up, and ran.

Aside from the problems of avoiding bees and finding monkeys, it had become clear to Tim that even sorting out something seemingly so simple as "feeding behavior" would present its own serious difficulty. Colobus monkeys are adapted, with chambered stomachs and specialized saliva and gut bacteria, to digesting leaves with a high cellulose content and low nutrition, but did they prefer mature or immature leaves? What else did they eat? Shoots? Flowers? Fruits? Ripe fruits or unripe? During which times of the year? From which trees? He had been trained in bird ecology, so he was aware of the scientific literature on birds suggesting that the nature of a species' food supply could correlate with the nature of its social system. Perhaps this relationship would hold true with the monkeys as well, but in order to follow that line of thinking he would also need to learn about things like the relative abundance and distribution and seasonal stability of the monkeys' preferred foods.

But first he had to recognize the trees. He contacted Steve Stephenson, chief warden of Gombe, who sent a first-rate botanist from the Tanzania Parks Department to come out and teach Tim how to identify trees. The pair of them began gathering samples of leaves and flowers and fruits from the many different tree species, drying and pressing them properly, and then sending them off to Nairobi for identification by the experts. Collection itself was a big task, and with the wetness and humidity at Gombe, everything would rot if you failed to dry it thoroughly. They were drying plant samples over an open fire, and then the samples would catch fire . . .

8.

Geza received an emphatically *final* notice from his draft board instructing him to report without fail on the morning of March 20 to the U.S.

Ruth, Jane, and Lori help celebrate Grub's second birthday.

Armed Forces Examining Station at 39 Whitehall Street in New York City. Around the middle of February, therefore, he and Ruth left for an extended holiday. Nic delivered the two of them to Kigoma. They took the train to Dar es Salaam, where they stayed for two days and two nights: time to enjoy their first experience swimming and snorkeling in the coral reefs at the edge of the Indian Ocean.

Then they caught a bus to the town of Arusha, in northern Tanzania, where Hugo met them and drove them north for a brief, touristic experience at Olduvai Gorge. At the edge of the gorge, Hugo stepped back to take a photograph: of Geza and Ruth standing there while dry, dusty winds tore at their clothes and hair. The wind plastered Ruth's hair, long and sun-bleached, over her right shoulder. Standing pensively in the stark sunlight and dressed in a light-colored, flower-print, one-piece combination of blouse and culottes, she carried a sweater casually in her left hand. Geza, wearing light slacks and a broad-checked shirt, sleeves rolled up high, stood and pointed at some interesting feature in the far distance.

From Olduvai, Hugo took them on to his and Jane's latest Serengeti camp—eight green tents arranged in a crescent on a minor hill overlooking a shallow alkaline lake with flamingoes pinking up the edges— where he was searching for wild dogs, the second carnivorous species of the *Innocent Killers* book. Hugo had been trying to find wild dogs for some time, in fact, but without success, and now he and Jane were about to pack up and return, along with two-year-old Grub, to their home in Limuru. Geza and Ruth went with them.

Soon it was mid-March. Geza and Ruth spent their final moments together before the drive to the airport. After that, Ruth stayed for another week and a half with Jane, Hugo, and Grub, along with another long-term American volunteer bound for the chimpanzee work. Loretta (Lori) Baldwin had arrived on the same Pan American flight that Geza left on, and Ruth and Lori were now together waiting for the small plane that would finally, on March 28, fly them both out to Kigoma.

It was a quiet time, those days of waiting, during which Ruth and Jane got to know each other better. And now that they had time to compare notes in depth and talk at length about the chimps, Jane discovered that she was indeed very fond of Ruth.

The most enduring remnants from that brief, bright sliver of time are three black-and-white photographs. Two clarify that Jane's mother had recently arrived from England, come to visit her grandson and help with babysitting. The third shows three younger women—Ruth, Jane, and Lori—standing and engaged in keeping two, probably three balloons aloft. Sitting in a high chair at a table below the balloons, a towheaded two-year-old smiles serenely as he looks up at the drifting balloons and waits patiently for someone to cut the birthday cake in front of him.

V

Love, Chimpanzees, and Death

(March to July 1969)

1.

When Ruth returned to Gombe on March 28, she was, as she wrote to Geza in the States, "so happy and excited that I cried."

Hugh happened to be in camp just then, sitting by himself on the slope near Pan Palace. Ruth stowed her traveling pack, then rushed out and sat down several yards away from where Hugh sat. He casually looked in her direction and then, quite suddenly, seemed surprised to see her, the actual her. He did a classic double take, looked at her once, twice, and then he sat staring at her while his hair spiked up in an expression of excitement. The whole sequence was somewhat "subtle," Ruth acknowledged, but it was also "beautiful." It made her "so thrilled." It meant, she thought, that Hugh recognized her, remembered her as an individual.

Ruth soon moved from her old room in Pan Palace into the aluminum rondavel formerly inhabited by Geza. A couple of grass mats had been tossed onto the concrete floor. Now a small wooden table and chair were pushed to one side, a cot and its mosquito net placed in another, and an Aladdin kerosene lantern was hung from the ceiling. Her small traveling suitcase, empty now, she slid under the cot. She arranged on some rough wooden shelves her books, the tanzanite jewelry Geza had bought for her during their trip to the Indian Ocean, and her personal journal. A few of Geza's things were still there, which she now stowed alongside her own clothes and other possessions. And so there it was: a hut of one's own.

The aluminum walls and roof of the hut had recently acquired a thin film of green mold inside and out, however, which was one indicator of the still advancing rainy season. Another was the flourishing of nettles, making Gombe "quite horrible when you crawl through a patch, get them all over, but have to grin, bear it and keep going to keep up with the chimp that's completely out of sight when he's just 3 feet ahead of you!" The grass had lately grown to around ten feet high, and a more general expansion in vegetation meant that many of the old trails had vanished. "Following is, consequently, quite a farce. Yesterday I spent 4 solid hours on hands and knees trailing after Mike and was it ever sweltering; he took me through some of the thickest places I know. Today I had quite a miserable time with 3 failures: lost Leakey, lost Mike, then lost Figan!"

A return letter from Geza gave the good news that he had submitted to the U.S. Army's medical exam in New York and was rejected because of his exotropic left eye and failing vision in the right. Ruth was delighted to learn of the rejection, although she then went on to read that Geza's graduate department at Penn State had refused to give him permission to return to Africa. He was required to stay in Pennsylvania until he finished his master's thesis, and Geza had no idea how long that would take.

Meanwhile, as Ruth wrote back, she was now fully settled in and determined to remain in Africa and continue working on chimp research. Having finished her first year on general observations—doing the A- and B-records—she was now free to work on a research project of her own choosing. This might or might not lead to an academic degree at a university somewhere, but at the moment, she was not interested in the academic possibilities. She just wanted to continue being with the chimps. As she had already discussed with Jane, Ruth was now planning to study ranging behavior among the adult males. Where they went. How far. How often. Why. She hoped to study nine adult males, and she was already discovering an interesting tension among the four highest-ranking ones: Mike, Humphrey, Charlie, and Hugh.

. . .

Among the males, dominance rank was routinely expressed in actions, postures, and gestures. Mike, just then, remained at the top. He was the alpha, having acquired that enviable position through a winning combination of chutzpah and moxie. Back in 1963, when he was probably in

King Mike seated.

his midtwenties, Mike had hit upon the novel idea of using large empty kerosene tins, stolen from Jane's tent and the camp kitchen, to enhance his masculine display. The other chimps had never seen anything like it. Those strange objects flashed and bang-boomed when rolled, and Mike learned to roll them about marvelously by pushing and knocking them with his hands while racing around the camp clearing. He became master of the flash-racket-boom display, and over time this astonishing wizardry made him master over the other males. They were intimidated. Jane and Hugo stopped leaving empty kerosene tins around, so Mike began using other foreign objects: boxes, tables, chairs, tripods. By the time Jane and Hugo had tied down and locked up everything they could think of, it was too late. Mike's arcane powers and high status were already established in the minds of the others. Even without the tins and other objects, everyone acted deferential, bowed obsequiously, and moved out of the way when he approached; and thus he had gained first access to all the important things in a chimp economy: food, sex, and respect. He was like a mafia don or a medieval king. So Mike was clever. He was also comparatively benign, not a bully like some of the other alphas from before and after. By the spring of 1969, however, he was moving past his prime. With Mike getting older, the three other top males, so Ruth began to believe, had turned from loyal

Handsome Hugh, relaxed but alert.

subordinates into scheming ones. And who, she wondered, would finally topple the king?

Humphrey was probably the most obvious candidate. Humphrey, who may have been suffering from a chronic ear infection, had the odd habit of poking a finger or thumb into one of his ears. But he was, in any case, also a big, strong bully, intimidating and notorious for his brutal attacks on other chimps, especially the females. Indeed, Humphrey seemed to dislike females generally, including human ones. He often threw big rocks at them. It's true that his aim was bad, and he did have a playful side. He would sometimes try to get some of the females or the adolescent males to play. But because he was such a thug, his please-play-with-me approaches seemed to make the females and young males anxious and evasive, which in turn made him angry.

Charlie and Hugh were the other two likely candidates for next alpha, Ruth thought, and they were probably brothers. They were both confident, powerfully built individuals, and they often traveled together and would back each other up. That gave them a huge advantage. So Charlie and Hugh could win the crown together—but they probably wouldn't share it. Which one would finally come out on top? Charlie had always seemed like an impressively fearless chimp, but Ruth thought Hugh was actually the one to watch. Hugh was a prime male with a weight lifter's bulk and a gymnast's grace, and in fact Ruth had long favored Hugh above all. She simply liked him. She liked his looks. She liked his style and character: the aplomb, the quiet self-confidence. And Ruth, given her own passion and imagination, had few inhibitions about describing to Geza her personal feelings about Hugh. Writing on April 14, she announced that she had "begun my relationship with Hugh!" The expression was mildly ironic and deliberately ambiguous for dramatic effect, I believe, and she followed it up with a description of touching Hugh on his back. "Have you ever touched Hugh's shoulder blades?" she began. "You really should. It's one of the most delightful experiences I know. Upon first touch he's warm and soft, but then warm and hard, solid muscle all the way down his back."

. . .

Ruth's predictions about the teetering social hierarchy were confirmed by events that unfolded over the next several days. "I go out almost every day," she wrote to Geza on April 22, "and things are really happening!" For one thing, "CHARLIE BEAT UP HUMPHREY!!" That act confirmed her suspicion that Charlie and Hugh were jointly up to something.

The fight began with Charlie acting disrespectfully to Humphrey. He waved a palm frond in front of Humphrey's face. Humphrey acted as if it were nothing, but of course, it was perfectly insulting for someone to act that way. Then, and without warning, Hugh emerged from the vegetation walking upright and beating his chest like a gorilla in front of Humphrey. Charlie, perhaps made nervy by the dramatic arrival of his older brother, proceeded to attack Humphrey—but soon, and possibly unnerved by the enormity of what he had just done, Charlie ended the attack and ran off. Hugh took over what his younger brother had started, however, with Humphrey by then simply screaming in terror . . . But suddenly, as if the weather gods had decided to compete with this advertised battle of the Titans, a violent storm appeared: a rushing wind and then a series of lightning strikes, followed by a sudden downpour over everyone and everything. The fight was rained out.

Later that day, Mike, still nominally alpha, returned from somewhere else in the forest, having been away for a few days. When he showed up in an area near the edge of camp where some of the chimps had started nesting for the night, Hugh—already settled into his nest in a tree—looked down at Mike. Then he climbed out of the nest and down the tree, and walked right past the old alpha without acknowledging his presence. To Ruth, it looked like another intentional insult, and Mike appeared to think so as well. It had been years since Mike had used the flash-racket-boom tins to enhance his display, and so now he displayed in the more normal way, while Hugh, continuing to act as if Mike simply did not exist, casually walked away. A while later Hugh returned, and Mike, apparently outraged at the previous slights, displayed aggressively in Hugh's direction once again. That was it. Both males squared off. Their hair began to rise with the surge of adrenaline and a growing fury. They began walking toward each other.

It was a game of mutual intimidation, a showdown, high noon at the OK Corral, and at the last second Mike gave in. As Ruth described it: "Mike veered off and walked right out of camp!!!!!!!!"

Clearly, major realignments were on the verge of happening, and Ruth believed that all of the nine males would soon experience shifts in their relative status. But Hugh was still her primary focus and her favored candidate for next alpha. In fact, he was her favorite chimp above all, the one she was now beginning (as she wrote on April 22) "to love . . . more than ever." She had that same day been alone with Hugh in the ravine beneath the Peak. He, sitting in a tree, looked down at her and then beckoned and shook a branch at her several times.

Beckoning and branch-shaking, and the accompanying swish-swish-swish of leaves, are aspects of a male-to-female communication among the Gombe chimps, with him indicating his fervent desire either to have sex right there or, alternatively, to go away with her to a more obscure place so that the two of them might be alone for a few hours or days and have sex without interruption. Ruth was amazed by the communication.

2.

After receiving a pair of affectionate letters from Geza, Ruth wrote back on May 5, declaring that she could find "few words to express what that first letter meant to me." She continued: "I have been indulging in a few pleasures lately, such as watching the sunset from a favorite tree on the beach ridge." She had taken his letter into the tree, and "the combination of joys" produced by reading his words and watching a sunset from the tree had been "quite something. It's an extraordinary thing to find all at the same time a person, a way of life and a means of learning that one loves with all one's being. You see, I too learn things in your absence and also yearn for your body."

But the intensity of her attachment to Geza seemed just then to be matched by the intensity of events happening at Gombe. For starters, the place had recently been shaken by an earthquake that appeared as a series of waves. The ground wriggled like jelly, which Ruth found astonishing. And later, while she was out with a group of chimps near the Kasekela waterfall, some of them feeding up in an enormous tree, she began to hear a rumbling beneath her feet that at first she thought she imagined. But the low-frequency rumbles heralded additional tremors, and soon the enormous tree began to shiver, to shift, and then slowly to topple, groaning and crackling until it hit the ground with a dramatic crash. The chimps who had been feeding there, even the ones in the tree, seemed hardly to notice what happened. They quickly recovered and just kept on eating.

Then there was the encounter with a mysterious serpent who, when she first noticed him, reared up like a stick come alive and gazed at her with gemstone eyes. For a while, a few minutes perhaps, Ruth and the snake took the measure of one another. She had no idea what kind of snake this was. She twice tried stepping closer, and the creature, instead of dropping back to the ground and slipping away, spread open a cobra's hood and hissed. That was, she wrote, "very exciting!!" She later described the incident to Nic, who identified the snake as a spitting cobra, and told her that spitting cobras spray a venom from forward-facing holes in their fangs. They direct the spray with impressive accuracy at any glittering object, such as someone's eyes, and the venom can cause permanent blindness. "It gives me such a satisfied feeling," Ruth concluded, "i.e., another day of danger survived!"

Meanwhile, on May 13, Hugh beckoned and branch-waved again, and this time when she didn't respond positively to his gesture, he displayed at her, then wrapped a hand around her ankle and dragged her along the ground for about fifteen feet. It was "a long way to be dragged and I can't say it exactly tickled," Ruth wrote. She was left with a large dark bruise on one leg, along with a scattering of cuts and scratches, but no obvious regrets. She must have understood that Hugh, big and powerful enough to have torn her to pieces in a matter of seconds, was altogether amazingly restrained—as if he had only momentarily been confused about whether she was another chimp or just another one of those faint and fragile chimp-watchers, and quickly regretted his own impetuosity.

. . .

Such adventures marked Ruth's quotidian experience, but now that her communications with Geza were reduced to the actions of ten tense digits laboring to activate three times as many keys on a mindless machine, now that she no longer had the actual Geza to confide in and be insulated by, she began to turn inward; and her increasing attachment to the chimps, and to Hugh in particular, was counterbalanced by an increasing detachment from the people around her.

It was an emotional problem, one can imagine, but what's the difference between emotion and biochemistry? After Geza had left Africa in March, she stopped taking her birth control pills, which at the time relied on much higher doses of estrogen than they do now, and the sudden cessation of that daily dose, with the resulting alteration in hormone levels, may have affected her mood. A possibly related matter: her menstruation cycle did not return to normal, so perhaps Ruth was also worried about being pregnant. But the social tensions in camp were certainly a major source of stress as well. With Carole down in Nyasanga, Ruth had no one she could talk to about the chimps. Tim Ransom was focused on baboons. Patrick McGinnis had gone off to do his academic term at Cambridge. Cathy Clarke had been there only since mid-October, Lori Baldwin since late March; neither of them, in any case, knew all that much about the chimpanzees. And what of the two Englishmen from Cambridge University who had showed up that January?

Michael Simpson, the senior scientist (and in Ruth's angry and obviously unfair assessment, "the biggest idiot I have ever known"), had recently been given a holiday with Jane and Hugo in Nairobi in order to reassess his role at Gombe. He was now back in camp, no longer a senior scientist and, generously, gamely, planning to help with the chimp observations and starting to learn the chimps' names.

Tim Clutton-Brock, the other new recruit from Cambridge, was still working on his red colobus study. He had by then learned the names of the trees in Kahama Valley, and he had also estimated the numbers of different tree species and their distribution patterns. He then etched a grid into the Kahama forest.

Since the red colobus monkeys ranged over a comparatively large territory in Kahama, roughly a square kilometer, that grid was necessarily large. He hired panga-wielding workers to chop transects: straight paths in the vegetation. The first set of transects ran parallel to each other, about one hundred meters apart, and on a compass bearing from the lake due east most of the way up the escarpment. Then he had the workers cut another set of parallel transects one hundred meters apart that ran north to south and thus perpendicular to the first set. Finally, then, he had created a matrix of hundred-meter squares in the forest, which meant he could walk around easily and have a decent measure of distance, direction, and location. He could also follow the monkeys without making a great disturbance below them, and one positive result was that they got used to him.

It made sense scientifically, and Tim had properly consulted the person from Tanzanian Parks in charge of Gombe, Steve Stephenson. The chief warden gave his permission—although Jane, according to Tim's recollection years later, "wasn't very happy about it" when she later learned what had been done. Ruth, meanwhile, was outraged. When, having followed chimps south all the way to Kahama Valley on May 13, she saw the cutting for herself, Ruth felt, as she soon wrote to Geza, "positively sick about it."

> Probably you were here when C-B was making plans for sectioning the valley? Well, now it is done and what used to be a place of beauty is now a mess. Literally every other tree is slobbered with thick white paint; one comes across a large plum tree with 2-foot-high [lettering]: T 953. I am immensely depressed by people who have no respect for anything and must leave their dirty smudge on everything they touch. And this is done in the name of science! But this is only a forest—it's too awful to think of all that has been done to animals and people under this same absurd justification.

"I'm beginning to withdraw from the people around here," Ruth admitted in the same letter. "Dinner conversations are nothing but recounts of who did what to whom all day, described in the most superficial and anthropomorphic manner. It makes me rather sick so I seldom go to dinner anymore." Altogether, in fact, she had not been feeling like her old self for a while. As she had mentioned to Geza on May 5, she seemed to be "in a very peculiar state of mind."

A compounding event, or trauma, happened around the same time when she received a letter informing her that her best friend from high school (someone she regarded as a "brother") had been killed in Vietnam. That information was passed on to Geza in the form of a handwritten note on a scrap of paper that was tucked into an envelope along with one of her typewritten letters. The note said, "Do you remember the one I used to refer to as my brother? He is dead now—killed in an ugly way in Viet Nam." His death, Ruth continued, was "bitter," and she was only telling Geza about it "because you mean so much to me." But, she asked in conclusion, "please never, ever mention it to anyone."

. . .

Then she stopped speaking to the others in camp. Carole had been unaware of this latest development until Nic brought Cathy Clark in the boat down to Nyasanga to talk to her. Perhaps Carole, as one of the old-time chimp people who had known Ruth and Geza during the past year, might be able to break through the barrier Ruth had erected between herself and the others.

Carole reacted to the news with disbelief, and Nic and Cathy stayed for an hour before they were able to convey the depth of their concerns about Ruth and convince Carole to see what she could do. A day or two later, Carole walked up along the beach, and when she reached Kakombe, she went directly upstairs to the main camp to find Ruth. It happened that Ruth was the only one there, and she was doing general observations. Carole said, trying to maintain a cheerful voice and demeanor, "Hi Ruth. How are you?"

No answer.

Carole said, "How's it been in the forest lately? Seen anything exciting with the chimps since I was last here?"

Ruth, in a cold, remote voice, scowled and said, "*What do you mean?*"

Carole was stunned. She wondered where he friend had gone. Why did Ruth suddenly hate her? She had never before seen Ruth that way, and she was shocked and intimidated. Ruth was twenty-six years old, while Carole was only twenty-one, and so her immediate reaction was defensive and deferential. She responded weakly: "I don't know. You know, just something that you . . . that you saw the chimps do."

Ruth didn't answer, and Carole didn't know what else to say. Ruth had cut off the conversation. Carole lacked the maturity or grace to break through the silence and address Ruth forthrightly, to say something like: "Ruth, is something wrong? What's bothering you? Would you like to talk about it? We used to be friends. I don't understand why you find my question odd. You know I'm very interested in chimps. I've always been, and we've shared that in the past." Eventually, after several minutes of facing what seemed like a crystal wall of hostile silence, Carole began to feel angry. She thought Ruth was acting superior, and so she became defensive. Later on, she wrote angrily in her journal: "I will never speak to her again."

Eventually, though, Cathy Clarke managed to penetrate Ruth's silence. Carole would years later recall the story for me as she understood it: that Cathy came to Ruth and said something like, "Ruth, you hurt me when you just turn your back on me and walk away!" She burst into tears when she said it. Ruth said, "Have you ever had a brother killed in Vietnam?" The two of them began to cry together, then to talk.

Finally, and in response to Cathy's urging, Ruth took the train east to Dar es Salaam to consult a doctor about her emotional state. The doctor agreed that perhaps going off birth control pills so abruptly was indeed the source of her problem. He told her, as she reported to Geza in a letter of May 22, that the "silly pill (or should I say the lack of it)" had left her feeling "deranged." The good news was that "two injections in the backside" had set her "back on the road to sanity." It was such a relief to find someone at once sensitive and authoritative who could tell her that "all the crazy, mixed-up torment in my mind" was the result, after all, of an ordinary hormonal imbalance. She expected to return to Gombe within a few days "in a much better and healthier state of mind."

3.

The rains continued through April and well into May. Then the sky began to open up, and Carole thought the dry season had finally arrived. At last! Then in late May came another pounding rainstorm, and after that she asked Nic when he thought the rains would stop. Nic asked the question of Rashidi Kikwale, who was the head of staff. Rashidi was tall and dignified, and he wore a knitted white fez. Carole was there when Nic asked him: "When are the rains going to stop?" Rashidi looked at the sky for a long while, maybe five minutes, as if considering the question with the deepest meteorological interest before answering, with what seemed like absolute certainty, "It'll rain once more." And he was right. It rained once more. After that, the weather shifted decisively, and with the end of May came the end of the rains.

Carole had by then begun finding some of the southern chimps every other day, but she was so tired that finding them was no longer enough. She was not encouraged by her first taste of success. And even though the dry season had arrived, she was also depressed and crying a lot. Yes, she was starting to find them, but now she was finding them to the south of Nyasanga Valley. South past Kalande and into the valleys of Kitwe and Gombe. So she had to cross a couple of valleys even to get there, and although the valleys were narrow and low compared to the northern valleys, getting there still took a lot of climbing.

In addition, she was getting weaker. She had learned to compensate for her increasing weakness by honing her skills at passing through the forest: what she thought of as her forest craft. She learned to see trails, to see openings, and to keep from struggling to go in a straight line. And she gave up smoking. She took the canister of Ujiji grass over to the choo and poured the sweet contents down the sour hole, so that was the end of that. Cigarettes were a different story, because she was really addicted to nicotine, but she quit smoking cigarettes next. Quit cold. She told herself that she would give up anything if it meant she could stay at Gombe. She didn't need the cigarettes, and they were draining her energy.

Tim Ransom finished his baboon study that May and returned to California. Once Tim left, Nic was Carole's only reliable friend. Nic was lovely, and since he and Margaret had broken up back in April, with Margaret flying back to her family in South Africa, things between Carole and Nic ought to have been better. They weren't. There were so many things she couldn't discuss with him in the way she had been able to with Tim. She started making tactless comments like: "Well, Nic, it's hard to talk to you. You aren't mature." He was six or seven years older than she, and he didn't know how to respond. Carole knew she was losing him.

. . .

After Ruth came back from Dar es Salaam on the final day of May, she seemed much better. She smiled. She greeted people. She was still quiet and self-contained, but she showed up for dinner and didn't seem hostile or antisocial. Carole, when she heard about this, and then when she went up to have a meal with the others and saw Ruth, felt relieved and pleased. Ruth was back to her old self.

Carole had never considered Ruth a close friend, and perhaps such a friendship wasn't possible as long as Geza was around. Nevertheless, she and Ruth always had in common their love of the chimps and the forest, as well as the experience of being coworkers and fellow B-record followers. And after Ruth came back from consulting the doctor in Dar es Salaam, Carole decided to try talking directly to her once again. She walked up to Kakombe one afternoon and found her, as before, by herself upstairs. Carole wanted to apologize for being unable to provide help when Ruth needed it, but she didn't feel comfortable about making the apology. She decided instead to talk about her own problems and ask for suggestions.

First, she told Ruth about how tired she was and how the chimps were always far away and never in the same place, which meant endless slogging through obstructive vegetation. What did Ruth think? Ruth suggested that Carole hire someone to cut a few footpaths, making it easier for her to travel. That made sense. That was a good idea.

Carole also said she was no longer sure how to approach the chimps because the valleys to the south of Nyasanga had less forest. In Nyasanga Valley, where there was still dense forest, she was usually able to get close to them through skillful creeping. Then, when she reached a point where she knew the chimps were going to see her soon, she would pretend to be digging in the ground or chewing on something delicious, acting as if she were preoccupied with her own important tasks and had no interest in them at all. It was a way, she thought, of demonstrating that she was not a threat. But now that she was going farther south, the openness of the vegetation—rolling grasslands with some woodland and gallery forests—often meant she could no longer creep up. What did Ruth think about that? Ruth said maybe Carole should stop trying the stealthy approach and be satisfied with showing herself out in the open.

Another thing that bothered Carole was the lack of a high vantage point. Down south there was no high vantage point like the Peak in Kakombe, where you could use binoculars to scan across long distances for indications of chimp activity. Ruth told Carole that she should hire some of the fishermen to build her a tree house or a high section of scaffolding in a tree.

So Ruth had some good ideas, and Carole wanted to try them. All along, however, she had been hoping that some of her chimp friends from the research camp would come down and interact with the stranger chimps where she was. That never happened, and Carole became more and more exhausted until, by the middle of June, a week or two after talking to Ruth, she finally decided to leave Gombe.

She went up to speak with Nic about her decision. Nic had been acting withdrawn lately, but when Carole said she wanted to leave, he worked hard to dissuade her. He said, "You've now begun to succeed!" And yet she was discouraged. She talked and cried for hours one evening. He listened patiently, tried to be encouraging, and in the end they wound up in bed. Then they had another argument that turned into a typical withdrawal on Nic's part—"I have got *nothing* to say"— and Carole was left with a terrible ache inside, a knot of painful feelings and the bitter knowledge that she was driving him away.

Nevertheless, Nic hired some of the fishermen to clear trails in places where Carole thought they could help her move from one valley to the next. He walked with her as she showed him where those trails would be useful. He marked them out, and the day he did Nic actually saw some chimps; and he knew that she was now finding them every other day. He had put up with her moods, her crying, and it was simply characteristic of him to encourage her now that she had started finding them. Finally, his encouragement made Carole feel better, and she said she would stay at Nyasanga.

. . .

At around this same time, Nic had a couple of unfortunate encounters with snakes. The first one was no more than eight to ten inches long with a big broad head and yellow spots on a green body, and Nic was trying to photograph the little reptile. This was one of the kind he thought could be a new species or genus that might be named after himself: a variety of bush or tree viper he envisioned as *Atheris pickfordi*. He stuck a stick into a container of sand, and he posed the snake on the stick. He was trying to get a good picture, so, with a careless reach of the hand, he adjusted the stick and *wham!* The snake got him. He was sore. The bite probably wouldn't have been fatal, but he went off to Kabanga Hospital, about seventy miles beyond Kigoma, for three days of being spoiled by the nuns of the Medical Missionaries of Mary.

The next time he got bitten was also a result of his own carelessness. The previous summer he had built a photographic darkroom out of cement and stones with a tin roof. He went into the darkroom to get something out of a box. He opened the box and a little green snake fell down from a rafter above and into the box. Nic knew of a little green snake at Gombe called *Philothamnus*, which is harmless and common. He said to himself, *Oh, green snake.* He picked up the snake and *wham! Oh! Juvenile black mamba!* Exactly the same color, but the head was a completely different shape. This one had the coffin-shaped head of a black mamba. So Nic went back to the same hospital where one of the sisters, this wonderful nun, said, "You're making a habit of this." Indeed, the second time was much more serious. His whole body hurt, the pain lasted for quite a while, and Nic stayed there for four or five days before returning to Gombe.

Hugo, as soon as he heard about the black mamba bite, wrote to Nic on June 26, instructing him emphatically to "please be careful!!" and reminding him that "in case of any serious accident there should be no hesitation in chartering an aircraft from Nairobi to fly someone to a good hospital. There is no need to delay this by waiting for an okay from me."

It was good of Hugo to remind Nic about the policy for emergencies, since Nic by then had begun to recognize that the Gombe bank account was virtually empty, and that Jane and Hugo had been paying for things out of their own personal accounts, which were now approaching empty as well. Jane had visited Gombe for two or three weeks at the start of the year. Hugo had flown to Gombe for a brief stay in late March. Jane had also planned to check on Ruth and the others in June. But then, after hearing that Ruth felt much better after her trip to Dar, Jane and Hugo jointly decided that they could not afford the nonemergency flight from Nairobi. There was hardly enough left to justify a short trip to Europe that had been planned for many months: for Hugo to attend the wedding of his brother, Godi; for Grub to visit his family in England; for Jane to attend three scientific conferences. They would leave on July 6, as Hugo informed Nic in the same letter. Hugo planned to return three weeks later, on July 25, to be followed soon after by Jane and Grub.

. . .

Meanwhile, Ruth's chimp follows had slowed down, mainly because there were fewer adult males around. Indeed, by the start of the second week of June, several of them—including her favorite, Hugh—had simply disappeared. Slipped away. Vanished. Gone somewhere else. Who knew where? Then, on June 11, Hugh reappeared. Wonderful! On that same day, Ruth was able to stay with him for about eight hours as he traveled along with Hugo and Figan. They went north as far as the high slopes of Linda Valley, and altogether it had been, as Ruth wrote, "quite a nice day." That was an understatement. It had been a splendid day.

Her stated interest was the males' ranging behavior, and by then, she was mapping travel routes taken by various males and was starting to discover some intriguing differences and similarities in their ranging decisions. Mike, for example, had the habit of going just so far up the Kakombe Valley before he would turn around, come back to camp, and then head out in a northwesterly direction. He did this even when rich fruits were in season only a couple of hundred feet past his turning-around place. Hugo, by contrast, would unhesitatingly go much farther up Kakombe Valley. But hardly ever, possibly never, would he journey far to the northwest. Meanwhile, Hugh and Mike never seemed to travel together except when they were both part of a much larger group.

Those were bits and pieces of a puzzle, and Ruth had been coming across other pieces of other puzzles as well, including one that would not be fully recognized for a few more years, which was the puzzle of chimpanzee warfare. She may have been the first to begin seeing and thinking about it, albeit uncertainly. During June and into July, she noticed that the males were sometimes going on patrol-like expeditions away from what seemed to be their familiar, core territory, which was centered on Kakombe Valley; and during those times they occasionally came into contact with individuals or small groups of stranger chimps. She had seen no predations in quite some time, and yet three times, by the end of June, she had witnessed something amazingly like the start of one, except that it would have been a predation on other chimps. She saw stalking, definitely stalking: chimps stalking other chimps, the behavioral prelude to a lethal attack. "It's quite interesting," she wrote, "but I haven't yet been able to figure it out."

But the pleasure she derived from being among the chimpanzees, feeling attached, feeling almost as if she were one of them, began increasingly to contrast, as it had earlier that spring, with an equivalently compelling alienation from the people around her. She could never enjoy socializing with her fellow *Homo sapiens* in camp. The two Cambridge men, for instance, seemed to Ruth to display the intellectual skills of their discipline in the style of a couple of displaying male chimps. It made her feel defensive, as if she were being told that she wasn't up to their standards, that she wasn't capable of being a *real* scientist. None of this may have been deliberate or even actual on their part, of course, but Ruth perceived it that way; and she believed, as she wrote to Geza on June 11, that Michael "looks down his nose at me because I don't count how many figs Mike eats relative to how many figs Figan eats. He spends ridiculous . . . amounts of time translating his notes into abstruse code."

For reasons that were "obscure," she confessed to Geza in a letter dated July 4, she had been going through "rather long periods of depression since you left (at least partly due to the hormone mix-up)," and that state of mind, she declared, was undoubtedly affecting her social connection with the others. As for the two academic Englishmen, how could she judge them? She couldn't. She did, though, once again decide to avoid associating with them. "The endless stream of pseudo-scientific clichés . . . annoy me no end, so I choose to be in their company as little as possible and things work out pretty well." She had lately decided that she would straighten herself out. She could find no rational reason for feeling anything other than blissfully happy, and yet it was also the case that straightening herself out required cutting herself off from the others and "going deeper into my shell." She was still showing up for dinner every night, placing her body in that circumstance among the others, but in her mind she floated away, stayed separate and private, especially when it came to news about the chimps.

Ruth saw that if she allowed herself to be concerned about all the things she might find distressing, she became resentful, which made personal relationships in camp utterly intolerable. So she was now determined to stop being concerned. She would keep to herself and allow all of the old issues and problems to pass her by. Just one thing really mattered, as she wrote with a surprising ferocity in that same July 4 letter: "The only thing that I care about is the welfare of the chimps, and when that is in danger, I will fight until death (literally!)."

. . .

Carole strongly believed that the experience of being at Gombe would make a positive difference in people. She saw a positive difference in Nic, for example. Even the fact that he fell in love with her was evidence of that. But gradually, during her time in Nyasanga, she, too—like Ruth—became disconnected emotionally and socially from the others. She didn't see them on a daily basis, after all, and she did not participate in the regular dinnertime conversations. She had started to become an outsider herself, and yet she maintained a strong sense that she wanted to be *as human a human as a chimp is a chimp.* That was a phrase she sometimes repeated to herself, and it meant that she as a human needed the society of other humans just as much as the chimpanzees needed, desired, loved, craved, and thrived on the society of other chimpanzees. Thinking that way, acknowledging that fundamental need for social contact, she became determined to visit the main camp as often as she reasonably could. She came up again on July 4, and Cathy said, "What are you doing here?" It seemed as if Cathy would say something to that effect whenever Carole came up, and she always took it the wrong way. It made her self-conscious. It hurt a little.

She said, "I just came up to see people."

She ate dinner with the rest of them, but it did not go well. She went back down to Nyasanga, returning to visit the main camp again a week later, on July 11. Early that evening she had an unproductive, unhappy talk with Nic, who still seemed remote, and although she stayed in the recently erected guest hut above the beach and just a short distance from the Cage, where Nic now lived, he and she did not share a bed that night.

On the morning of July 12, which was Saturday, Carole woke up early and listened for footsteps, hoping that Nic would at least come over to say hello. He didn't. She heard him walking about. Then she heard him get in the boat and start the motor. He was going into town for supplies. And Carole, conscious of Nic's withdrawal and aching with that knowledge, slept late. At last she woke up and thought about getting up to urinate, but she didn't want to move. In fact, she couldn't move. She was unable to persuade her body to sit up and climb out of bed or do anything else. She had a fever, and she was extremely weak. She just lay there. Tried to sit up a second time. Couldn't. So she remained as she was until, later in the morning, Jumanne Mkukwe appeared, and she told him she was *mgonjwa*. She was sick, probably with malaria, and she asked him to run upstairs and tell the others.

Michael came down with a thermometer and quinine. He took her temperature. It was 104. He gave her the quinine. Then she was in bed, lying in the guest hut with the day's heat pressing down heavily while a dozen restlessly peregrinating flies above her spun their angular fantasies into the air. That's when Ruth disappeared. Ruth followed Mike out of camp at around noon that day, and she did not return.

4.

At first, Carole didn't know that Ruth had disappeared. The malaria left her only half aware, drifting as she was between a leaden wakefulness and a febrile skein of bad dreams and strange thoughts. Ruth's disappearance became apparent to the others at dinner time. Nic and Patrick (back by then from his first term at Cambridge), along with Cathy, Lori, and Michael, ate together that night, and when Ruth didn't show up, someone went upstairs to check her hut. Not there.

Tim Clutton-Brock missed dinner as well that evening. He had recently built, on the ridge between Mkenke and Kahama Valleys, a simple hut— grass walls, tin roof—for overnight stays, so that he could get to his red colobus monkeys early in the morning when they started their day. Michael walked there and found Tim in his hut, informed him that Ruth was missing, and the two of them returned to the ridge camp together. Tim told the others that he had heard boisterous chimp sounds coming from upper Mkenke Valley that afternoon, between three and seven, when he was out watching his colobus monkeys. He said he had seen neither Ruth nor chimps, though, and had not heard any human cries for help.

It was dark by then, and everyone decided it was time to start looking for her. Given Tim's recollection of noisy chimps in upper Mkenke, combined with the fact that Ruth was last seen leaving camp following Mike and heading south, they all decided to concentrate on an area to the south of camp, although they also knew that Mike could easily have changed directions, turned west and gone higher or veered north.

The six of them split into pairs—Nic with Cathy, Pat with Lori, and Michael with Tim—and began the search. It was clear that Ruth could be lost somewhere in an enormous and rugged area. Perhaps she had been bitten by a snake. Maybe she had broken a leg or been injured by a buffalo. Or was she stricken with malaria and unable to move very far? The field packs they were all supposed to carry included whistles, matches, flashlights, flares, and snakebite kits. But what if her pack was lost or incomplete? What if the flashlight batteries were dead or the flares soggy? It was also clear that they needed to find her right away. Because it was night, they stayed on established trails and were careful to illuminate the paths before them with their flashlights, watching out for snakes and listening for other potential dangers. They called and called: "Ruth! Ruth! Ruth!"

Nic and Cathy walked the ridge between Sleeping Buffalo and Kakombe all the way to the tree line at the top, investigated a couple of the *korongos*—ravines—that were part of the Kakombe Valley, and then returned to the main camp to check once more at Ruth's hut before hiking up to the Dell and on to the Kakombe waterfall: calling and calling, stroking and poking the darkness with their flashlight beams. Pat and Lori walked up to the high ridge of Sleeping Buffalo, while Michael and Tim passed along the high Mkenke transects before meeting up with Pat and Lori on the Sleeping Buffalo path. And Tim, upon returning to his overnight hut on the Mkenke-Kahama ridge, called out into the upper Mkenke area and Kahama Valley.

At around one o'clock in the morning, Nic walked down to the guest hut by the beach to tell Carole that Ruth was missing. He said that Dominic would be taking care of her, and he instructed her not to get out of bed for three days.

. . .

The next day was Sunday, July 13. By then, everyone understood the absolute seriousness of what had happened. They had to stop what they were doing and find Ruth. With everyone getting up before dawn and starting at first light, they began the search in earnest. Nic and all the researchers except Carole were now joined by all available men from the staff. Dominic and Sadiki stayed back to run the camp and organize the meals, but everyone else joined in. That included Rashidi Kikwale, of course, as head of staff, and also Hilali Matama, who was in charge of the field staff. The field staff had by then been expanded to include

Hamisi Mkono, Eslom Mpongo, Jumanne Mkukwe, Yusufu Mvruganyi, and one or two others. Ferdinand Umpono, the Parks Department ranger, came up from Nyasanga to help as well.

Everyone spent the day going north, covering Linda Valley, then sweeping south again all the way to Mkenke. But they also searched the area around camp more thoroughly with the advantage of daylight. That afternoon, Nic and Pat took the boat into Kigoma to report Ruth's disappearance to the police. They returned that evening followed by a police launch containing what was called an emergency field force that included twenty-two officers altogether.

With the police contributing, the searchers were able, by the start of the second full day, which was Monday, July 14, to work in long lines of about forty people altogether, using walkie-talkies to keep in touch and combing through specific areas methodically. On Tuesday, July 15, a plane was flown to Kigoma from the Parks Department headquarters in Arusha, and Patrick took the boat down to Kigoma to meet the pilot and assist him in an aerial search that finally began after the winds settled down that afternoon. Patrick also, while in town, sent a telegram to Louis Leakey in Nairobi: "PHONE JANE THAT RUTH HAS BEEN LOST FOR 3 days." Louis immediately telephoned Jane and Hugo, who were then in Holland attending the wedding of Hugo's brother. They relayed the news by telegram to Geza in Pennsylvania and Ruth's parents in Lynchburg, Virginia. For the next four days, the parents would be in daily telephone contact with Hugo, who was in turn communicating daily with Louis Leakey.

Nic had been coming down to the guest hut by the beach each day to give Carole the latest news, the theme of which was invariably the same, with minor variations: "No, we haven't found her yet. We're going to go again tomorrow." Late Tuesday night, however, Carole was instructed by Ruth herself to get up from her malarial bed and contribute to the search. Carole had been sleeping deeply, she told me, when the sound of footsteps crunching the gravel outside her hut woke her up. After the crunching stopped, the door flew open with a bang, and there was Ruth, standing right there and covered with blood. She seemed angry or aggrieved, and she said in a fierce, demanding voice: "*Why haven't you come to find me?*"

Carole, terrified by the vision and voice, sat upright. Having gone through more than three full days of the quinine treatment, she felt stronger and not delirious or confused. She was thinking clearly, and she thought: *This is real.* But once she sat upright like that, Ruth was replaced by darkness. Still, the vision had seemed completely coherent, and Carole decided it was a communication. She got up the next morning while it was still dark, pulled herself out of bed, put on her normal clothes, and went upstairs to join in the search. The others tried to discourage her, saying things like, "No, we don't want you to. You could have a relapse of malaria, and then we'd have to look for you. It would make things worse. You should just go back to bed." Nic was the most emphatic of them all. But she said, "No, I've had a dream that she came, covered in blood, and said, 'Why haven't you come to find me?' And so I'm going to look. There is no way you can stop me."

During the first night of the search and the day that followed, people were thinking they had to find Ruth quickly. As the days passed, though, that sense of urgency diminished, and the searchers began to dread what they might find. How long could anyone survive out there with a broken leg or a poisonous snake bite or some other kind of disaster that made her unable to move or to call out? So when Carole joined the others on the fourth day, which was Wednesday, July 16, people were feeling less hopeful and more desperate.

That afternoon, Gombe warden Steve Stephenson arrived in the trimaran, bringing along a small contingent of extra game rangers to join up with Ferdinand Umpono. And on the fifth day, Thursday, July 17, around sixty of the Gombe fishermen joined in the search.

Geza was by then staying with Ruth's parents and her two sisters in the Davis family home in Lynchburg, Virginia, waiting with desperate hope for some good news. He knew from reading Ruth's letters that she had recently followed chimps into the Kahama Valley. Kahama was a long and difficult trek away: two valleys to the south of the research camp in Kakombe Valley. No one from among the searching wazungu imagined that Ruth had gone that far following a chimpanzee during a single afternoon, which is apparently why the search, as it continued through days one, two, three, four, and five, amounted to a series of increasingly refined forays around the main camp valley and the immediate valleys to either side. But then Geza sent a phone message from Virginia, via Hugo in Europe, to Louis Leakey in Nairobi saying that they should try looking in Kahama Valley. Carole, since she had done the long B-record follows during the summer of 1968, remembered one time following Mike and a group of others all the way to Kahama; she also knew that Ruth had disappeared following Mike. So when an abbreviated version of Geza's message was finally relayed to someone at Ramji Dharsi's store in Kigoma, and from there to someone at camp by late Thursday or early Friday, Carole decided that she would, she *must,* go into Kahama and look for Ruth there.

. . .

Friday, July 18, was the last day of the search. Steve Stephenson had decided that Ruth could not possibly have survived that long. He announced to everyone that Friday was the last day, saying, "There's no point in searching for a person who will have been dead and has already been eaten by scavengers." Stephenson didn't know the ecology of Gombe very well, though, and he may have been wrong about the efficiency of the scavengers there, but when he said he was calling off the search, Carole's spirits sank, although she was still determined to go to Kahama.

The sixty fishermen who had participated in the search the day before had done so at a sacrifice. It was dagaa time, with the moon closing into a silver sliver, and so the fishermen were replaced by a large contingent of children from Mwamgongo who had been let out of school in order to help. Thus, the sixth and final day of the search was marked by the chirpy, excited voices of several dozen schoolchildren released temporarily from school. A few teachers and school administrators had also come along to try keeping the youthful chaos under control.

Patrick and Michael went north to make a final pass through Linda Valley, while Nic, Carole, Cathy, Tim, Lori, the African staff, and the cheerful mobs of Mwamgongo schoolchildren walked in a long line up the Mkenke Watu path. They intended to form a united front and, using whistles to keep contact with one another, to sweep south along transects leading in the direction of Kahama, then regroup farther down the slope and sweep north back into Mkenke. But while they were all still climbing up the Mkenke Watu path in order to execute that plan, Carole took Nic aside and announced that she was going to explore Kahama by herself.

Nic looked at her angrily. Once he recognized her absolute determination, however, he deferred, albeit with a compromise. He said, "I'm going to send you with a bodyguard."

Nic asked Ferdinand to go with her. He knew Ferdinand was the ranger Carole had been living next to down in Nyasanga, of course, and it was a good choice. He instructed Ferdinand emphatically in Swahili. Although Carole didn't know Swahili, Nic spoke slowly and clearly enough that she could still figure out what he said, which was something like this: "Don't look for the missing memsahib. You keep your eye on this memsahib, and you make sure you're able to take us to her if anything happens to her. You are responsible." Ferdinand was a competent and conscientious person, and he listened carefully to Nic's instructions and would, Carole knew, take them seriously.

The two of them, Carole and Ferdinand, went on ahead, farther up the Mkenke Watu path, and as soon as they were out of sight of the main search group, she felt better. In fact, her B-record follow of Mike and four other male chimps a year earlier had also gone up the Mkenke Watu path before cutting to the south on an animal trail. She was worried, unsure that she would be able to find the animal trail Mike had used a year ago. But she knew there would be a crossing somewhere, an intersection of animal and people routes, and she was trying to figure out exactly where it was. She wanted to be intuitive about it, to reach into her memory and her intuitive sense, and so they proceeded up the Mkenke Watu path a little farther, until she came to an area where something told her: *This feels about right.* She said to Ferdinand, mixing English and Swahili with gestures to make herself clear: "Now I want to look for animal trails." They began to look and then saw one, and Carole said, "Yes, I think this is right." So they turned off the Watu path and began to follow the animal trail.

By then they had gone above the forest line and were moving into a steep area of dry grass and scattered trees, and now, following the clues of that faint track—Ferdinand staying ahead, his panga in hand, Carole walking behind—they began traversing the headwall of Kahama. As they continued along, they came across a scattering of dung and fruit pits, the remnants of old *mbula* and yellow cluster berries. Recognizing them as chimp leavings, they stopped to examine them more fully. Carole spoke to Ferdinand in another broken Swahili exercise: "Looks about six days old." Ferdinand nodded in agreement. And they went on. It was a long walk, and Carole was just trying to remember the trail she had taken a year ago, when she had followed Mike and the four other chimps. Obviously, she could never be certain that Mike would have taken the same trail almost a week ago, but it was the best route she could think of. Now, as she began to discover signs of chimps, she began to think, *I might really find her!* That thought led to a painful spin of emotions. Terror. Fear. Grief. Hope. She wanted so badly to find Ruth alive, wanted to do everything in her power to make her be alive. She began to feel a faint hope at the prospect, and the hope helped for a while. They continued on, still finding indicators of chimps having recently passed through, and finally they came around to the southern side of high Kahama. There they found a dry korongo marking one of the high tributaries of Kahama Stream: a deep and ragged ravine filled with rocks and cliffs where the water had gone underground. Carole recognized it from before, and she knew now, knew for certain, that they were passing along the same trail and going into the same place she had gone with Mike and the four other males a year earlier.

Carole said to Ferdinand: "Let's stop and look right here." The animal path went onward, but she wanted to climb down into the korongo for a moment and peer beneath the cliffs and under the vines growing around the boulders. She did that. Ferdinand stood and watched while Carole went down and investigated those dark interstices. She started calling, "Ruth! Ruth!" She was hoping so hard, hoping she would hear a weak answer from someone who had, perhaps, broken a leg, and as she began to recognize that no one was there, she started to cry. But she kept looking. She crawled on her hands and knees, checking everywhere along the base of all these great boulders, looking beneath the vines, into the shadows, until finally she stopped, wiped the tears away from her face, and said to Ferdinand: "Let's go down the stream and see if there are more cliffs." He agreed. So they walked farther along the bottom of the dry korongo until they reached a spring and the start of a stream, and then they followed the stream. They had been in the dry grassland with a scattering of trees, but now they were moving into more lush vegetation that got thicker, and they could see the stream.

They sat down, listened to the splashing sounds of monkeys leaping through leaves, the murmur of flowing water, the calls of hidden birds, and the endless buzz saw of invisible cicadas. And they noted that the ground and the stream flowing along seemed to have leveled out. They walked another seventy-five yards or so, walking right in the water, following the stream, and the stream was flowing smoothly. Carole thought they must have gone below the cliff line and had reached the level floor of lower Kahama Valley. They stopped. Carole said to Ferdinand: "Maybe there are no more big rocks." She didn't know the Swahili for *cliff* so she just talked about "big rocks." He nodded. The stream-edge gallery of vegetation was thick there, and Ferdinand had his panga out, ready to clear the way a little to make the going easier for Carole, who had been crying and was still weak from malaria.

According to Carole's forty-year-old memory of that day, the two of them stood there talking about "big rocks," and she thought they had come to the end of them: with the stream flowing coolly around their feet and past them. They must have reached the valley floor, she thought. But then Ferdinand stopped suddenly. In Swahili, he said, "There's a big rock right here." And when he said that, Carole's perception of a continuous valley floor dissolved and resolved into a second perception, in which the clear water, flowing so coolly past their feet, did not continue meandering along in front of them but instead gathered itself and dropped. And the thick gallery of bushes and underbrush around them, instead of continuing before them as she had previously perceived, merged into the tops of high trees that were lining a valley below. They had walked up to the top of a waterfall, and even the sound of falling water had been masked by the closer, more immediate sounds of moving water and their own splashing feet. Swelling clear and cool, the stream slipped through the vegetation, passed visibly along until it slid into a moment of shadowed invisibility where the vegetation closed in. The stream swelled quickly there, and the waterfall began. Carole started to step forward, closer to the brink in order to look over. Ferdinand put his arm in front of her and said, "No. You must not."

He indicated that they should go around the falls, find a route that took them around so that they could make their way safely down to the base. It took them fifteen minutes to a half hour, with Ferdinand leading the way, Carole following behind. And now, once again, she compared where they were in reality to her memory of following Mike and the four other chimps on that day during the previous summer, and she realized that the chimps had led her around the steep part and the waterfall. They had taken her through the forest and down to the lower valley floor, and they had done that well enough that she never knew the cliff or the waterfall existed. She never saw the waterfall, never knew about the cliff, even though she had gone past them.

She and Ferdinand reached the bottom of the cliff and turned to follow the stream back up toward the pool at the base of the waterfall. Walking on the valley floor now—he proceeding about six feet in front of her, hacking away at the occasional obstructing vine or branch with his panga—they moved through a forest that was green and rich, with great trees reaching high above them and a gently wavering bank of vegetative understory below, a glowing puzzle with filaments of light passing gloriously through a wall of leaves, and then Carole saw the waterfall. It was enchanting, and Carole looked up in awe at a dappled beauty that was just so fresh. She had never seen it before, that particular waterfall, and she always loved the beauty of Gombe. Everywhere you looked was beautiful—but suddenly Ferdinand caught his breath and pointed. "Huko!" he said. "There!"

She stepped up beside him and peered through a veil to see what he was talking about. At the edge of the shallow pool at the base of the waterfall, on the stones there with the cool water flowing around and beneath: a bloated body. It was Ruth, her body swollen enough to have ripped her shirt partly open. Carole looked at the horror for a brief moment and started saying, "No. No. No. No."

She ran away, while Ferdinand stood there longer, looking carefully. They had both been around twenty or thirty feet away from the body in the water, and Carole just ran back and grabbed a tree and hung on. She just hung on. She couldn't stand up on her own. She tried, but she couldn't stand up, so she just hung on and kept saying, "No, no, no, no, no, no."

Ferdinand came back and said, "Give me the walkie-talkie."

She gave it to him. He tried to make it work. Couldn't. Carole's eyes were streaming tears, and she didn't know much about the walkie-talkie. Ferdinand finally gave up on it, and he said, "I'll run up. You stay here." He saw that Carole was clutching the tree because otherwise she was going to fall down, so he started to leave; but then Carole said, "No, I'm coming too. I won't stay here."

She was starting to calm down, and now the place where she was, that exquisitely beautiful place of a minute ago, was permeated with a rank horror. She began shivering. Ferdinand must have thought he was going to run fast and Carole wouldn't be able to keep up with him, but she insisted on coming with him. She couldn't stay there. Ferdinand began to run, and Carole was running after, slipping, stumbling, desperately trying to keep up.

After a time, they stopped to catch their breath. She had her small day pack, which Ferdinand now took and carried himself. He began to run again but had to keep stopping to wait for her to catch up. They passed out of the thickest part of the forest and moved higher into grassland and finally up to the Mkenke Watu path, where Carole just collapsed and cried. Ferdinand pulled out his ranger's whistle and began blowing on it. He blew and blew until it seemed as if he would burst, and the sound reverberated among the hills and valleys and into the korongos. Carole began to hear voices and shouts, faint and indistinct at first, from down below in Mkenke; and she listened to a lilting transmission, one voice passing information on to the next, relaying information back and forth in a lovely series of floating calls that traveled down into the dark emerald valley of Mkenke and back up again. She heard, then, one voice clearer than the others, saying or shouting something in Swahili. She didn't know what the voice was saying, and she sat there on the edge of the Watu path, the tears pouring out of her eyes and streaming down her face. Finally, one of the extra game scouts showed up, and after Ferdinand told him they had found Ruth's body, he turned around and ran back down to fetch the others. Soon more game scouts appeared, followed by some of the camp staff, and someone from the staff ran back to get Rashidi. Rashidi then began shouting for the others.

Nic had been down below with Tim when Ferdinand started blowing his whistle, and at first Nic thought the sound was coming from the north and that it was one of the Mwangongo schoolmasters trying to round up the children. But when the whistling continued, he knew it was from the south and that something was wrong. Then, when he heard Rashidi's voice shouting, he was certain that something serious had happened. Rashidi would not have shouted otherwise.

Nic and Tim ran all the way up the Watu path until they reached Carole and Ferdinand. They were soon joined by the game scouts and the African staff. Nic spoke to Ferdinand, who described what had happened, and then Nic came over and squatted down, settled a hand on Carole's shoulder, and said, "You can go home, Carole. We'll do it. Ferdinand will show us where to go. So you can go home." Then he spoke to Rashidi, and said, "Please, will you take her home?" Rashidi reached out with his hand to take her pack, and then the two of them began walking.

. . .

Nic and Tim, the game scouts, and some of the Gombe staff followed Ferdinand south to Kahama and then down to the pool at the base of the waterfall and Ruth's body. The head had been smashed open and was now a mass of maggots. One of the Africans made a brush of leaves, gently brushed away the maggots. Then someone made a litter of poles and vines, and Nic and Tim placed the body on it and covered it carefully with a blanket, and they all proceeded down the stream to the lakeshore.

At the same time, Rashidi and Carole were walking down the Watu path on the way to the main camp. About halfway down, though, they ran into Cathy and Lori, along with a large group of schoolchildren. Carole was crying so much she could barely see. Cathy immediately came over and hugged her, and Carole hugged Cathy back, and they wept together. Lori was there, too, in tears and patting Carole, the three of them standing in a huddle with Rashidi standing to one side.

Cathy said, "Is she dead?"

Carole said, "Yes." But then her mind began to work, and she began to remember that she had just seen the body for a few seconds, then run away. Maybe she was mistaken. Perhaps Ruth was still alive, and Carole had been too shattered and weak to understand that. So then she said, "I don't really know. I didn't go close and look. I'm not sure." And that was her delusion, her faulty apperception created by the force of her wish that Ruth was still alive, and it was unfortunate that Carole said it, she thought, because now Cathy had to endure that uncertainty as well. But Cathy didn't respond to the comment. Instead, she said, "Maybe Rashidi wants to go and help the other men." And to Carole, she said, "We can walk down with you, Carole."

So Carole said to Rashidi, "The memsahibs will come with me."

And Rashidi said, "No, Mama. I'm coming with you."

Mama! Carole had never before been called *Mama* by anyone. The Africans usually called white women *memsahib* when they intended to be respectful, but, of course, that word also had a history and colonial reverberations. It was what servants said. *Mama* was a warmer and less complicated expression of acceptance and respect ordinarily reserved for mature African women, not young wazungu, and Carole knew instantly that Rashidi meant what he said. It made her feel better, and she said to Cathy, "No, Rashidi's coming. He's taking me home."

The four of them walked back together down to the main camp and, from there, down to the beach and the guest hut and Cage, and they waited in the Cage. Three hours later, the police boat came around the headland and anchored a little off shore. Nic climbed off the boat and waded to shore. Carole could see he was troubled. He sat down inside the Cage, unlaced his boots, took them off to pour out the water, and said, "They insist on your coming in with us, Carole." Nic had tried to persuade the police otherwise, he went on to say, but they wanted her to go to Kigoma and formally identify the body. He added, "You'll have to get ready. Get some clothes. You'll have to stay a couple of days."

She ran to the guest hut where she had spent the last few days, gathered some of her things, then ran off to the choo. She could hear Nic yelling, "Carole! Where's Carole?" Then, when he saw her emerge from the choo, he said, "Oh," and he hoisted himself aboard the police launch. Pat was next, clambering aboard in the waist-high water, and Carole ran out into the clear water and stepped up after him. Soaking wet, she stumbled from stern to bow, passing on the way several policemen and a blanket-covered heap that smelled obscene. The three of them stayed at the bow during the trip to Kigoma. Nic, sitting next to Carole, refused to look at her for the entire time, and she didn't know whether his ignoring her was from shock and anguish or the fact that they had not been getting along.

VI

Aftermath

(July 1969 to 2007)

1.

The gray VW bus was parked in the usual place down at the Kigoma harbor, and Nic drove Carole and Patrick in the bus to where they were all staying, which was the house of the captain of the *Liemba*—the giant steel boat built by Germans during colonial times a half century earlier that still managed to split the waters of Lake Tanganyika, delivering people and goods from one end to the other. The captain was just then out on the lake with the *Liemba,* but his wife, Margaret, was home. Nic, Carole, and Patrick left their overnight bags at the house and then drove on to the Kigoma Club and got drunk. As they staggered out later than evening, Nic affectionately draped his arm around Carole. It felt good, and she recognized that he had softened up a little. Perhaps, she thought, there was still hope for the relationship.

Early the next morning Pat walked over to Ramji Dharsi's store on Lumumba Street to use the telephone in order to call Louis Leakey in Nairobi and inform him that Ruth's body had been found. Pat lifted the receiver, depressed the button three times, and a woman's voice spoke. While he chatted with the operator about getting through to Nairobi, Carole and Nic went off to the hospital to identify the body.

. . .

The hospital was a concrete building, and they were shown into a room with a concrete floor, a table in the middle, and a body beneath a sheet

on the table. The weather was very warm, and the place had no refrig-
eration, which meant that the body's decomposition was accelerating. A
man stood in front of the sheet-covered body on the table, drew back
the sheet, and Carole saw maggots and hair. But it was Ruth's hair. Carole nodded, her vision turning into a wash of tears. The man silently
pulled the sheet back into place.

It was still early, and Carole and Nic next walked over to the police
station to contribute to the police report. Nic did most of the talking.
Carole was still crying. But like all the other chimpanzee researchers,
Ruth had carried with her a tape recorder to speak notes into, and
although the recorder had been destroyed, the tape was miraculously
intact. At the police station, they listened to the tape as it turned on a
new machine provided by Ramji Dharsi. The sound was not easy to
make out. Ruth never said the date, but they heard "down into
Kahama," which was enough to presume it described her last follow.
The police said the tape should be transcribed, and Nic took Carole
back to the house, where she sat down in front of a typewriter.

Carole wanted to do this. She wanted to hear from Ruth. She wanted
to listen to her voice, know what happened, and she spent the next few
hours carefully transcribing the tape. She was seated at a table in front
of an open window, typing on an old typewriter, multiple sheets of thin
white paper interspersed with black carbon sheets and the recorder off
to one side. She played back the tape while a warm morning breeze
passed through the window. White curtains lifted and fell in the breeze;
and Carole sat there with the fresh, moving air caressing her face and
hands, and typed out the final words from Ruth. The tape smelled
faintly of death, but Carole was getting used to the smell.

She typed from the start of the tape, where Ruth said she left camp at
12:32 P.M. following Mike and headed south. Ruth lost him right away,
but she heard chimp hoots coming from high on Sleeping Buffalo, and
when she got there, she heard more hoots from Upper Mkenke. At 2:54
she reached the Mkenke Watu path and discovered a group of chimps
there. Upon seeing her, they immediately broke up and headed south
into Kahama. Mike wasn't in that group. It was Faben, Figan, Sniff,
Godi, Willy-Wally, Hugo, Hugh, and Charlie. A big group of males.
Then Ruth changed the B-record target individual to Charlie. There was
nothing unusual about the record. There were comments on feeding,
grooming, a large bird passing overhead, the sound of a boat motor
down on the lake. The chimps seemed to be steadily moving, while
Ruth was clearly getting tired, gasping for breath, her voice ragged and

rasping as she spoke. "So . . . ahhh . . . 3:14 . . . ahhh . . . Hugh catches up with . . ." She was laboring to keep up with the chimps and out of breath, but it was all normal. She had come far, and the chimps had been moving fast. Good observations. It made perfect sense. It was a chimp record. There was nothing depressed or despairing or suicidal in it. There was no tone of voice or mood filtering through to suggest that her thoughts were anywhere other than with the chimps and the sunny day and the long follow. She finally said, "I've lost them as they . . . coming down a dry stream gully."

That was the end of the record. The time: around 3:45 p.m. And Carole was certain that the "dry stream gully" in the tape was the same place she and Ferdinand had come upon the day before—and also the same place where, the previous summer, Mike and the other four chimps had led Carole into Kahama. But unlike Ruth, she had managed to stay with the group, and they had led her right up to and around the water-fall, whereas Ruth, having lost them, must have followed the stream and walked right off the leading edge of the falls. It was a great relief for Carole to realize, after she had transcribed that tape, that Ruth had been herself right up to the end. She was following the chimps she loved. She was tired. She came to a place that was not easy to see. She lost her balance and fell, and in falling she lost her life.

. . .

Not long after Carole had begun transcribing the tape, Patrick showed up at the house, reporting that he had not been able to reach Leakey in Nairobi, although he did leave the message that Ruth's body had been recovered. Patrick and Nic then went down to the harbor, where there was a boathouse and workshop. Nic said to the man there, "I need a coffin, and it's got to be watertight."

They found a carpenter to make a wooden coffin and a cross, and they hired a couple of metalworkers to fabricate an aluminum lining for the coffin. After that, they walked over to the town hardware store to look for coffin handles and something that might be used to seal the aluminum lining. The only sealant they could find was window putty, which they bought, along with a half dozen chromium-plated towel racks, short ones, to serve as handles. Pat then returned to Dharsi's store to try calling Louis Leakey again while Nic returned to the cap-tain's house to find Carole. As soon as Carole finished her typing, she and Nic took the typed transcripts off to the police, and then they sat down at the station separately, each to give a final statement for the police record.

Patrick showed up at the police station some time later, looking for Nic and Carole. He said that the phone call to Leakey in Nairobi could not be put through until late that afternoon. Nic had finished his statement to the police by then, so he and Patrick left Carole and walked back down to the harbor to check on the metalwork and screw the towel-rack handles onto the wooden shell of the coffin.

Then Ferdinand Umpono appeared at the police station, having just been brought into Kigoma from Gombe so that he, too, could make a statement about what happened. He passed on to Carole the news that Dr. Leakey was right that moment on the phone at Ramji Dharsi's, asking to speak to Patrick. Since Patrick was down at the harbor, Carole rushed over to the store and spoke to Louis Leakey on her own. It was by then four thirty in the afternoon. He said he could find no airline willing to fly a decomposing body anywhere, so he thought it would have to be cremated first. He also wanted to know if Ruth had died immediately. Carole said she had. Leakey said he would personally relay that potentially comforting information to Ruth's family and also to Geza. Finally, he said he wanted Patrick to call him back.

Carole returned to the police station. Not long after she had finished her statement, Nic and Patrick showed up. Nic said that the metalworkers down at the harbor had been working on the coffin lining for the last eight hours and still weren't done. Patrick said he had managed to reach Louis Leakey, who told him that it was not possible to get the body cremated in Tanzania, probably owing to Muslim traditions forbidding it. Thus, it had been decided that Ruth would be taken back to Gombe and buried there. It was highly unusual for someone, particularly a foreigner, to be buried in a national park, but Louis arranged it. Ruth's parents had already agreed to the idea. It did seem like the best solution.

The metalworkers were finished by nine o'clock that evening, whereupon Nic and Patrick dropped the aluminum lining into the wooden coffin, and then they hauled it up to the hospital in order to put Ruth's body inside. The body was far gone by then, and neither one of them could stand it. The sight was horrendous, the smell worse. They couldn't do it, and eventually they paid a couple of orderlies to place her into the coffin. After that was done, the coffin was brought back outside. Ruth was under a blanket, but the maggots were crawling out from beneath the blanket. Nic and Pat lined the upper edge of the aluminum basin with a continuous strip of putty, then placed the aluminum lid on and screwed it down tight. Then they screwed on the wooden top and brought the coffin down to a cool and secure spot at the harbor.

. . .

Meanwhile, Margaret, the captain's wife, had taken Carole out to see a movie. Once a week a movie would be shown at some place in Kigoma, and this time it was at the Kigoma Club. It was an old film, a blood-and-guts adventure story about the French Foreign Legion called *Beau Geste*. Carole wasn't enjoying it, and at intermission time Nic and Pat showed up wanting a couple of drinks. They had planned to go back to the captain's house and take baths before heading off to bed, but instead they decided to go out to the Kigoma Club. They were joking with each other, a desperate kind of catch-and-throw repartee where every remark produced a laugh and another joke, and soon they were drinking as fast as they could manage. Carole and Margaret left the movie and went into the bar to sit with Nic and Pat.

Nic, sloshing down double and triple whiskeys, was the drunker of the two, which is why he gave Patrick his car keys. At some point, Pat stood up and announced, "I'm going home." He staggered out and drove the VW bus back to the captain's house. Then, at around twelve thirty, Margaret, who had young two children back at home, left—after locating a Belgian friend in the bar, Marcel, who promised to take Nic and Carole home when they were ready. So Carole stayed with Nic. He seemed in a cheerful sort of drunken state, and she felt a warm connection with him. He was no longer avoiding her. He was looking at her, offering her cigarettes, and after Carole stumbled over to visit the choo and then came out again, she found him standing outside waiting for her. "Hello, Carole," he said. And then: "The way I see it, she's only matter. It's only matter now."

They tripped their way back to the bar and continued drinking until Nic went back to the choo and didn't come out. Carole eventually went to look for him and found him on the floor vomiting, unable to stand, so drunk he couldn't see. He just lay on the floor and vomited for another half hour. Then he passed out, lying against Carole and snoring. After a time she tried to rouse him, pull him upright. "You can't do it alone, Carole," he murmured. "You'll have to get someone to help you." He flopped down like a dead fish—Carole gently inserting her satchel under his head to keep it off the wet, stinky floor—closed his eyes, and started snoring again.

Carole went into the bar, found Marcel, and together they drew Nic upright. They pulled and pushed him out to Marcel's vehicle, levered him onto the back seat, stuffed his feet in. With Marcel driving, Carole in the back trying to keep Nic from flopping onto the floor, they made it back to the captain's house, where they dragged Nic into his room and dropped him as gently as they could onto the bed. Marcel left. Margaret helped Carole undress Nic. They left him there with, next to the bed, a bowl to vomit in.

. . .

Carole had a terrific hangover the next day. Pat wouldn't admit to one. Nic wasn't in the mood to talk about anything. In the afternoon, he and Carole returned to Gombe in the Parks Department's *Triton* in order to begin preparations for the burial.

Nic and Carole told the other researchers that Ruth would be buried at Gombe. They were, according to Carole's recollection, "shocked and angry" to hear the news. They had endured so much already, and for the nightmare to continue was hard to bear. But Nic soon found a beautiful spot for the grave: on a grassy ridge just above the beach, beneath a *myombo* tree between Kakombe and Mkenke Streams, and with a good view of the lake. Then Nic arranged with Rashidi to make sure someone dug the hole. Other preparations included fixing up a place for Ruth's parents to stay, since they were going to fly out. And getting a place ready for Hugo, since he was on his way back to Africa and would be escorting the parents from Nairobi to Gombe. In addition, a place had to be found for Ruth—her coffined body—to be kept overnight before the burial. In fact, there were many things to think about, since a number of people from Kigoma would be coming in for the ceremony, but what number? And would they expect to be fed? Who would officiate? What kind of priest or minister could they hope to get? What was Ruth's religion anyway? Did she even have one?

Nic went back to Kigoma late that afternoon to rendezvous with Patrick and begin addressing some of those questions and issues. Carole, now at Gombe by herself, slept down at the beach in the Cage. I will imagine that Carole, restless and alone that night, missing Nic and feeling tired beyond reason but kept awake by grief and anguish, stepped out into the darkness and looked up. When the sky is clear of clouds at Gombe, it is possible to see a universe alive with stars. With only minimal spill from the flux and dross of civilization's electromagnetic radiation in the visible wavelength—from house lights, streetlights, automobile headlights, and so on—the sky takes on a weighted and mystical presence. Instead of squinting at a hundred or five hundred faint, flickering points of light, you gaze up at an entire sea of them, ten million or a billion distant suns swimming within galaxies and clusters of galaxies, at times brightened by a moon—which, on the evening of July 20, 1969, was a waxing crescent. If Carole did go outside, and if she looked up and had the requisite clarity of vision, perhaps assisted by a good telescope, she might have discovered a dark speck moving across that drifting slice of moon. The speck was the lunar module of the American *Apollo 11* flight, settling down at 20:18 Greenwich mean time onto a barren stretch known as the Sea of Tranquility. Inside the module sat two men, Neil Armstrong and Buzz Aldrin.

By the time Armstrong opened the hatch and stepped outside, leaving a footprint in the dust and uttering a few clever words memorized for the occasion, Carole would have been asleep. It was nearly morning in Tanzania. In Washington, DC, it was still evening, and when Ruth's parents settled into their room at the Jefferson Hotel, exhausted and depleted, hungry and beyond hope or consolation, they turned on the television to watch the same two men taking their first steps on the moon. Mr. and Mrs. Davis had already gotten their necessary shots, except the one for yellow fever, and they were expecting the State Department to issue emergency passports and visas in preparation for their flight the next morning to London and the start of their hard journey to Gombe.

Hugo had also returned to Africa by then, while Jane remained in Europe with Grub, and so it was Hugo who, on Wednesday, July 23, met the parents at the Nairobi airport after their flight from London landed at eight o'clock that morning. He greeted them, loaded their luggage into the back of the Land Rover, and then, after stopping at the American embassy for help in solving last-minute problems having to do return tickets and yellow fever shots, drove them out to Wilson Airfield on the southern edge of town. There they met the pilot, "a very nice young man," as Mrs. Davis characterized him in her journal. They clambered over the wing and crawled through the side door of a four-seater Piper Cub.

2.

After the funeral—with Ruth ceremonially blessed, lowered into the ground, and covered by a mound of earth and stone; with the hundred and more mourners of all colors and many cultures ferried from Gombe back to Mwamgongo and Kigoma and places in between; and with

Hugo flying in the Piper Cub back to Nairobi alongside Mr. and Mrs. Davis, who would then catch their return flight to America—Nic found to his great relief that there was less to think about and do. The only obligation remaining was the inquest in town, which he, Carole, and Pat attended on Monday.

The courthouse consisted of a brick wall about three feet high with pillars, a roof on top, and a completely enclosed office along one side. It was open on the other three sides, so that a refreshing breeze could pass through. At the inquest, a representative from the police gave his report, while the coroner read the results of his autopsy. Nic thought the autopsy was thorough and the report well-written. It covered everything, including *empty uterus*. Ruth had not been pregnant at the time of her death, and she had never been pregnant. That information was significant, as was the fact that her skull had been entirely split open, front to back, which confirmed that death had been instantaneous. And, the police concluded, no foul play had been involved.

The inquest was brief but as important as the funeral service, being another symbolic scrap of activity that brought substance and sense to a transubstantial and nonsensical tear in the world. And when Nic got back to camp, he pulled out the jacket he had bought for the funeral—a white, collarless Mao-style jacket of the sort favored by President Nyerere—and gave it to Rashidi. "Here you go, Rashidi. I don't think I'll be wearing this again." It had been the only dress jacket available in Kigoma at the time, and Nic was glad to be rid of it. He was also grateful to Rashidi for his calm help during a hard time, and he thought that the staff under Rashidi's direction had dug a beautiful grave. Nice and straight. They really put a lot of work into that grave.

Nic remained at Gombe for the rest of the summer, and then in September he went home to his parents on the farm in Kenya for a visit. When he returned in October, he discovered that the Gombe finances were a mess. It took him three twelve-hour days to straighten them out, and after that, and being fed up with Hugo's administration, he decided to leave. He left on November 13. He and Cathy Clark had become a couple by then, and after a while they both went to California and got married.

They enrolled together as students at the University of California at Riverside, but the marriage didn't last. They drifted apart, and he went to live with his brother in central California until he could afford a house of his own. Nic's brother had earlier left Kenya and settled down in California, and Nic, while living there with his brother, got a job on a large almond farm. He worked as a mechanic and tractor driver, and he sprayed the almonds. It was a lovely life. It didn't pay much, but it was fun, and he got to know America and Americans better.

Nic had liked Cathy a lot, but to him at least it was never completely clear why they got married. As the years passed, in fact, Nic came to believe that he did many things for reasons that were not completely clear. He didn't know himself very well in those days, and it took a long time before he realized that he was actually a very naive person. He trusted other people far more than he should have. But over the years of living in America, he saw a lot, and by the time I talked to him in 2008, Nic was starting to get the idea that he could tell, after meeting someone for the first time, which side of the line that person was on. Being able to read others: some are born that way, while others are not. He was not. He was born completely innocent. He believed everything, and it cost him a lot over the years.

. . .

After the funeral, Carole stayed at the main camp in Kakombe and began joining the rest of the wazungu for dinner, although it seemed to her as if they all became more indirect and thoughtful whenever she appeared, as if she weren't supposed to be there. Perhaps some of those feelings had to do with her own growing estrangement from Nic, who, she could see, was starting to develop an interest in Cathy. Yet Carole's sense of being an outsider was more than that, she told me in 2009. She had spent all those months down at Nyasanga while the others had been socializing together in the main camp. And then, she thought, the others, during that superheated time of stress and despair as they searched for Ruth, had bonded as a group, while for most of that time she was overcome by malaria and lying in bed down by the lake. So she would go to dinner, but she began to feel that people's words were saying one thing—welcoming her to dinner, expressing polite curiosity about how she was doing or feeling—while their bodies and faces said something else. Carole had learned to read body language from the best experts in the world, the chimpanzees, who compensated for their limited spectrum of vocal communication by being superb at reading gestures and postures and facial expressions. Carole had acquired some of that chimp social intelligence, and her chimp intelligence told her that she didn't belong.

She stopped going for drinks or dinner or other group events with the other wazungu. Instead, she would wander over to Dominic's kitchen or to his house. He and other people around would say, "Karibu, Mama

Carole! Welcome! Oh, come in! Sit down!" She felt she had risen in respect among the Africans, and now these were the only people who comforted her, and for a long time after that she trusted black people more than white people. She hung out with the Africans during the final weeks she was there because they were safe. She also went and sat with the chimps. But the Africans, both men and women, were warm and welcoming, and they would say, "Why are you leaving us? Why are you going?" And she would say, "I don't know. I don't understand how it all has happened the way it happened."

On the day she left, all the wazungu came down to the beach, and she took a picture of them. She wasn't in it because she took the picture. And there they stood, all with what seemed to her like mildly uncomfortable smiles, because they were probably glad to get rid of her. That's the way it was. At least that was Carole's perspective on the matter, and she was devastated by the idea that they all had not been made better by the influence of living close to nature and knowing the chimps. The whole experience shattered her deluded dream that Gombe would magically make people better human beings.

She went to Nairobi and stayed there for the next three months, doing her best to remain in Africa and see whether she could heal herself. She clung to the hope of returning to Gombe some day, and she visited Jane and Hugo briefly in their home during that time, wanting to explain her hope to them. Jane told her that she was welcome to come back at any time. Jane was, in fact, very sympathetic. She said, "I understand you found Ruth's body. That must have been pretty terrible." Carole had by then learned to tone back her boisterous American style, and so she didn't blurt out things the way she used to. She never told Jane the full story. She never said much about what it was like for her. She just said, "It was very, very hard."

No one talked about post–traumatic stress disorder in 1969. Maybe the term hadn't been invented then, even though all those Vietnam veterans were beginning to come back crippled with psychological injuries. But Carole concluded many years later that she must have experienced something like PTSD. She was scared a lot, on the verge of panic. People frightened her. She was unsure of herself, had trouble making decisions. She tried to make up her mind whether to leave Africa or not, and she couldn't. She had not wanted to leave Gombe, and she didn't want to leave Africa either. Gombe was the one place she wanted to be. But at the same time she was not healthy enough to be there. It would be dangerous to be there, in her depleted state. She was only twenty-one years old, and yet she hadn't much physical stamina left, so she stayed in Nairobi and tried to figure out what to do next.

She reconnected with a few of her Quaker school friends, although the main group had gone on to India, while the FWI African headquarters by then had been moved out of the city and into a suburb. She briefly tried to renew the old relationship with her Gangster Boyfriend, but she wasn't the romantic, heroic young woman she had been a year earlier. She was instead a needy, uncertain, traumatized person, and she didn't trust the Gangster Boyfriend as a friend. She was not ready to tell him the whole story of what happened. When she once tried to describe it, he didn't seem all that interested. She had been hoping for sympathy.

Another boy in Nairobi, Michael Levin, who was also an FWI student but staying in town on his own, was a more sensitive sort, and he became a friend. He listened to her story and was sympathetic. He introduced her to an African named Richard Lanoi and also to Hassad Said Mohammed, who was the game warden for Nairobi Park. Those three became her saviors. They comforted her and talked to her, heard her story and accepted her grief. But when she met other people in the bright streets of Nairobi, it was different. Someone would try to talk to her, say hello, and she couldn't say hello back. Her throat would constrict. Her mouth would dry up and she would be unable speak.

After three months in Nairobi, Carole got a job at a place in Cornwall, England, called the Monkey Sanctuary, which was a sanctuary for South American woolly monkeys. Working there was a terrific experience, although she kept having terrifying dreams about Ruth, and the woolly monkeys were not as exciting, socially or intellectually, as the chimpanzees had been. Chimpanzees will eat for six hours and still have another six hours left over for playing, relaxing, scheming, and socializing. The woolly monkeys were not like that. She stayed in England for about a year until, in 1971, she returned to America and headed to California, where after several years of uncertainly and struggle including, at the age of thirty, the diagnosis of Type I diabetes, she completed a master's degree in plant ecology and got a job working for the State of California as a plant ecologist.

It made sense. She loved being outdoors. It was the second best life she could imagine. The job lasted several years, and she was working on habitat restoration—but in spite of the pleasures of doing that, she always understood that she liked animals better than plants. Plants don't look you in the eye or get to know you. Finally, after her second mother died in 1995, she bought a ticket to Africa. She spent five years working at various jobs in five different East African countries, and then she was accepted as a volunteer doing baboon research in Mikumi National Park, Tanzania. Mikumi was about six hundred times bigger than Gombe, but it was still in Tanzania, still East Africa, and now she was told she could still do what she had done at Gombe, except that it would be more savanna than forest and baboons rather than chimpanzees.

The diabetes was always a serious concern, however. Before starting at Mikumi, Carole flew to South Africa to buy a years' worth of insulin and syringes and test strips, along with three blood glucose meters. In Tanzania, she bought glucose powder, and she packed the powder in film canisters. Then, once she began at Mikumi in the fall of 2000, she was prepared to deal with the diabetes. If her blood sugar level went low, she could open one of the film canisters and pour some glucose powder into her mouth. If it went high, she could give herself a shot of insulin. But even without those exceptional events, she was giving herself a minimum of four insulin shots a day, including one long-acting dose before she went to sleep and a short-acting one just before the bowl of oatmeal she ate for breakfast. After breakfast, she would pack up her day's supply of water, food, insulin, and glucose powder. Then, joining up with the ranger who always accompanied her, Carole made the ten-mile drive from camp out to the sleeping tree of the baboons and sat beneath that tree waiting for the baboons—feisty little guys with Swahili names like Amka, Heshima, Huruma, Mbilikimo, Pumzi, and Upende—to wake up, climb down, and start their day.

. . .

Back when she was at Gombe, Carole had stopped taking photographs, because what was the point? You had this flat little representation of what was in front of you, but the forest was everywhere. No photograph could begin to show it. So Carole had always wanted to thank Hugo for making that wonderful film about Gombe called *People of the Forest*. And after her work was finished at Mikumi in early October 2001, she was able to speak to Hugo in person about the film. He said that the editing alone took him five years, and it nearly bankrupted him. She said, "Well, you did make the most wonderful gift to the world with that film." She really meant it.

When Carole saw Hugo that time, he was old and crippled by emphysema, most likely caused by heavy smoking and the Serengeti dust. He had lived in his tented camp in the Serengeti for many years, photographing animals, but now he could no longer do that sort of photography or tolerate that environment. He was approaching the end of his life yet still smoking, only now he had to be satisfied with tiny puffs. And even though he and Jane had been divorced by then for nearly thirty years, Hugo went to Dar es Salaam and began living in the guesthouse of the seashore property Jane had inherited from her second husband after he died. Hugo had nowhere else to go. Carole was there only one afternoon, and she had come mainly to see Jane and get her permission to make a return visit to Gombe. After getting that permission, Carole got on the train the next day and rode west to Kigoma and Lake Tanganyika. She made it back to Gombe at last, and she had two weeks there.

Things had changed so much that it was frightening. Thousands or tens of thousands of people who had been displaced by wars and conflicts in neighboring states—Congo, Rwanda, and Burundi—were settled into large refugee camps around Kigoma, and all the trees that had once been visible outside the park's boundaries had been cut down. The land from Kigoma up to the park's southern boundary was now a barren sea of eroding brown hills, and, as Carole soon learned, the devastation affected not only the flora outside the park. The people camped there, poor and displaced, had begun setting snares inside the park.

At the same time, however, she could see that there were several promising changes. Perhaps the most fundamental one began as a direct result of Ruth's death. It would have happened eventually in any case, I believe, but Ruth's death prompted the new rule that any researcher going into the forest had to be accompanied by a field assistant. As a positive sequel to that new rule, the African field staff became professionalized more quickly and thoroughly than they might have otherwise. Within a few weeks or months, they knew the forest as well as any of the visiting researchers did. Within several months, everyone on the field staff had learned to carry out basic scientific observations on chimpanzee behavior. When Carole returned to Gombe in 2001, the research station was run significantly by Tanzanians, with an expanded staff of around twenty-five, and it was supporting first-rate international behavioral studies on chimpanzees and baboons, as well as several other species. There was plenty of good work going on, in short—even though, as Carole soon discovered, that expanding success was accompanied by increased restrictions for visitors. All visitors were required to stay at least twenty-five feet from any chimpanzee in order to avoid transmitting infections. Non-Tanzanian tourists were required to pay a hundred dollars per day to enter the park, plus more for food and lodging. They were also expected to hire field assistants to serve as guides.

A few of Carole's old friends and acquaintances from the staff were still there, but Rashidi Kikwale had died of malaria about ten years earlier. And hardly any of the chimpanzees Carole had known back then were still alive. Flo, Figan, and Faben were all gone, as was Geza's friend Leakey. The four males Ruth had favored—Humphrey, Mike, Charlie, and Hugh—were also gone. Mike died of pneumonia in 1975. The other three were probably killed during the intercommunity war that took place in the latter half of the 1970s. Nevertheless, Flo's daughter Fifi was alive and thriving, and she had by then produced eight offspring, including two powerful and socially ambitious males, Freud and Frodo. Both had risen to alpha status during the 1990s, Freud first in 1993 with his younger brother Frodo wresting away the crown four years later. Frodo, notorious as a large and unpredictable bully, still reigned as alpha when Carole visited—although he would soon, within a year, be overthrown during an extended illness by a rebellious coalition.

On her best day there, Carole had a chance to follow Melissa's daughter Gremlin. Gremlin had been born a few years after Carole left Gombe in 1969, and now she was a fully mature female, a mother with three children: eight-year-old daughter Gaia and a pair of five-year-old twins, Golden and Glitta. By good fortune, there was no one else that day, no human present other than Carole and the field guide who came along as her minder. They went out for three hours, and the experience was the closest she came to returning to the paradise in her memory. There was one lovely moment when they crossed a stream. Mama Gremlin carried one of the twins, Golden, and she crossed the stream by moving through the trees overhead. Eight-year-old Gaia was with the other twin, Glitta, not carrying her, and Gaia jumped across the stream. The little one behind her, her sister, whimpered and couldn't make it. So Gaia turned around. She was merely a youngster herself, but she reached back with her foot to bridge the gap, and little Glitta climbed onto Gaia's bridge and crossed the stream that way. It was a beautiful vignette of helpful family relationships, and Carole concluded that the older sister Gaia was part of the reason that the twins, Golden and Glitter, were still alive after five years. It was a sweet thing to see, Gaia being helpful like that, and Carole liked her a lot because of it.

3.

Geza had been in Pennsylvania on July 15, 1969, trying to make sense of the data from Gombe for his graduate work, when he received the telegram from Hugo in Holland informing him that Ruth had been missing for three days. He immediately drove down to Lynchburg, Virginia, to be with Ruth's parents and her two sisters. Her father was an engineer with General Electric, and, like Ruth, he was reserved, never one to talk or reveal much. Her mother worked in an office. They lived in a new, ranch-style house with landscaping, flowers, a ravine and creek out back, and Ruth's bedroom at the end of a hall. The room was sunny and feminine, with a fancy bedspread and pillow shams and ruffles on the lampshades.

It was an excruciating few days marked by bitter coffee and solemn words. Geza talked about what Gombe was like, what might have happened to Ruth, what probably had not happened. Everyone tried to act hopeful, and he felt beside himself with dread and guilt. He was the one who had brought Ruth to Africa, after all, and now that she was missing, or worse, his strong impulse was to fly back and find her himself. He should fly back, but how? He had no visa. Getting one would mean going back to Washington, and it would take days. He had no money. He had no ticket and no way of getting one. His savings had been used up paying his own return flight back to the States, and his graduate fellowship at Penn State would not begin paying anything at all until the fall term.

It seemed as if the phone was ringing all the time. There were regular calls from friends, associates, members of the extended family. There were daily long-distance calls from Hugo and Jane in Holland, and at five thirty in the morning on Saturday, July 19, a call from Western Union relayed the message from Louis Leakey in Nairobi that Ruth's body had been found and that she was going to be buried at Gombe.

Geza returned to Penn State and buried himself in his work. He felt numb. He didn't cry. He didn't talk. He couldn't think. Nothing made sense.

They had planned to be married at Gombe. They had talked about the idea to Jane back in March when he was getting ready to fly back to the States for his army physical. Jane had seemed thrilled by the romance of it and the novel idea of a wedding at Gombe among the chimpanzees. That was how they left it at first: they would get married at Gombe soon after he returned in June 1969. When the Penn State anthropology department informed him that he would be required to stay and finish his work using the data he had already acquired, they agreed to marry when he returned in January. And their letters to each other since he had left that March were warm, intimate, and loving—although sometimes strained by distance and the uncertainty that distance can entail.

After her death, he put away the letters and photographs and mementoes. He finished his work on the subject of chimpanzee predation at Gombe, which became his master's thesis, and when he became a PhD candidate at Penn State the following year, he decided to return to Gombe in order to gather new data for the big dissertation. It must have been a painful decision; but in the summer of 1970, he flew to East Africa, dropped in to visit with Jane and Hugo at their camp in the Serengeti for a couple of weeks, then flew on to Kigoma. He finally stepped onto the familiar pebbled shore on the warm afternoon of July 22, a day he would years later mistakenly remember as July 12, the first anniversary of Ruth's disappearance and death.

. . .

By the time Geza arrived at Gombe that summer, only two researchers were present who had known Ruth from the year before. They were Loretta Baldwin and Patrick McGinnis. Lori had taken over Carole's old job of trying to habituate the stranger chimps down south in Nyasanga Valley, so she was not often at the main camp. Pat took Geza to see Ruth's grave, which was a heap of stones marked by a weathered wooden cross, but visiting that grave was about the only contact the two of them had that recalled Ruth or implied she had once been alive at Gombe. Even in normal circumstances, Geza's general demeanor and self-possession could make people careful around him, and now, given the discipline with which he avoided talking about Ruth and kept to himself, no one voluntarily raised the subject. Not Pat. Not any of the African staff. Not Jane or Hugo.

Geza visited the aluminum rondavel where he had lived and which Ruth had taken over after he left. It was inhabited by another researcher now, someone young and new to Gombe, and not a single object remained in the hut that recalled Ruth or her work. All her things were gone—taken, so he imagined at the time, by her parents a year earlier when they came to Gombe for the funeral. Geza had left some of his own possessions in that rondavel with Ruth when he returned to America in March 1969, and those were gone as well. As for the most important object of all, which was Ruth's personal journal, he presumed that Mr. and Mrs. Davis had taken away that treasure as well. It was not in the hut.

He went up to the main camp to look through the filing cabinets inside Pan Palace. The narrative record, both the A- and B-records, had been discontinued immediately after Ruth's death in favor of timed behavioral checklists. But Geza soon found in the old A-record file a reference to Ruth in a footnote at the bottom of page 674, which was the final page. The footnote went like this: "July 12. 12:34. Ruth follows Mike out S. and does not return. We search this evening and until the 18th of July when we find her. Record A is discontinued." He could find no other mention of Ruth's disappearance anywhere, and since that terse footnote mentioned only that she had been found, without identifying the condition in which she was found, it is also the case that he discovered no reference anywhere in the files to her death.

It was strange, particularly in a place like Gombe, which, as he would later write, "was awash in paperwork about everything that happened every day to everyone working there year after year." But since he himself had begun working to forget, he was not sufficiently troubled by those minor signs of institutional amnesia to protest or to search further. Instead of fretting about such small mysteries, he did his best to concentrate on his work and following chimps. Oh, there was that new senior scientist from Amsterdam University who wanted to cut paths in Kakombe Valley for the sake of someone's convenience, and Geza felt that cutting those paths would invalidate his own data about chimp movements and preferences. He also saw the plan as a variant of the idea that doing science justifies casual destruction, so he and the senior scientist had an apparently dramatic confrontation over the matter. But generally Geza did his best to avoid drama. He stayed out of trouble, did his work, tried to fit in.

By September of that year, Jane had finished writing her popular book about Gombe, *In the Shadow of Man,* and it was with tremendous relief that she sent off the manuscript. In the same month, the book about East African carnivores that she and Hugo had labored on for so long, *Innocent Killers,* was released by the American publisher, and soon she was off to America to publicize that book in dozens of media interviews. She then traveled west to give several fund-raising lectures while also conferring with Professor David Hamburg at Stanford University. Within a year, she would be appointed a visiting associate professor at Stanford; but now, in the fall of 1970, she and Hamburg were actively developing their plans for the big alliance between Gombe and Stanford. Since the National Geographic Society was continuing to withdraw support, this promise of a major new partnership was welcome indeed, and when Jane returned to Gombe on the final day of 1970 and reviewed the latest research and the people involved in it, she was brimming with positive spirits. "Gombe has NEVER, NEVER been better," she began in an enthusiastic letter to her mother that ended with a bit of happy gossip: Geza and Lori Baldwin were "having a passionate affair!"

At that point, in fact, it may not have been so passionate or even much of an affair, but Geza's emerging association with Lori could still be regarded as a sign that he was dealing with his grief. He finished his study in the summer of 1971 and returned to Penn State. A few weeks later, Lori followed, and they became a couple in a relationship that lasted for about ten years. He completed the PhD in 1977 and soon turned from academics to conservation. During the late 1970s and into the 1980s, he and Lori worked together in Sierra Leone, West Africa, surveying chimpanzee populations, tracking the exports of live chimpanzee babies out of the country, and beginning the legal, political, and publicity battle that would ultimately put an end to that harmful business. And then, in 1986, he and Jane began working together on national and international matters having to do with chimpanzee conservation and care.

. . .

By 1989, as I mentioned at the start of this story, the three of us had decided to collaborate on a book about chimpanzees and their problematic relationship with humans. During the two or three years we three worked together, I never once heard Geza mention Ruth. I first heard the story of Ruth's death in a highly abbreviated fashion from Jane. If I had then thought more deeply about it, I might have imagined Ruth's death to be an unhappy event that was finished and fixed in time: insect in amber. I would have concluded that Geza had recovered from the trauma and gone on to become the person I then believed I saw before me: a self-confident, vigorous man fully engaged with the world. His successful relationship during the 1990s with Heather McGiffin, their happy marriage and family life, their talented son, their pleasant home in suburban Washington: all of that suggested to me a life undisturbed by regrets or ghosts from the past, the kind of life made whole and even possible by the gift of forgetting.

When Geza first telephoned me in 2006, saying, "Dale, I need to talk," he had stopped trying to forget and was working to remember.

4.

Remembering brought its own trauma, marked at first by the event of September 27, 2006, which I described in the opening chapter: Ruth's strange appearance one night in a fashion or form I referred to as "a vision or visitation or visit." Perhaps I should have chosen more boringly clinical terms. It was a haunting, true, but one that came entirely from the self. It was a psychological phenomenon, in other words, a vision or hallucination provoked by profoundly unresolved grief from the past combined with the present crisis of a body assaulted by poison and falling to pieces.

Haunted, in any case, by that resurgent reminder of someone who embodied a glorious time and place taken away so outrageously, Geza turned to the difficult task of recollecting and reconstructing a life experience that had, since the summer of 1969, been packed up and locked away. It was an intense burst of activity, that process of unpacking and unlocking, which finally produced five very large binders filled with old and new correspondence, written recollections, maps, photographs, and other materials.

Geza's immediate motivation for remembering may have been to counter the lingering rumor that Ruth had committed suicide, which he regarded as offensive and false beyond consideration. I've long admired Geza's thoroughness as a researcher, but I also understand that his version of events could never be considered perfectly complete or fully reliable. The past is a forbidden nation. We are all biased memoirists cursed with disintegrating memories. And so, for example, his insistence that Ruth could not possibly have jumped deliberately, seems to me an instance of opinion buttressed by emotion, a strong wish rather than a reasoned argument. Suicide cannot be so readily ruled out. It is reasonable to insist that any final theory about how she fell take into account her mental or emotional state.

For most of those who were involved in the immediate aftermath of Ruth's disappearance and death, including the researchers and the police and the experts who signed the autopsy report, the conclusion that Ruth had not committed suicide led directly to a second possibility: that it had been an accident. This second way of falling is what Carole strongly believed, and she described for me in dramatic detail her own memory of having walked up to within a few feet of the leading edge of the Kahama waterfall without—because of obscuring vegetation—recognizing the danger. My written reconstruction of the final day's search is largely based on a long personal letter Carole wrote to her friend Tim Ransom a few days after the discovery of Ruth's body, combined with Carole's own vivid recollections, told some forty years after the discovery. Those two sources, separated in origin by four decades, are not perfectly congruent. The early letter to Tim describes the daily progress of the search, but it says nothing about Carole almost tumbling over the falls herself because of obscuring vegetation at the top. For Geza that incongruence was a critical issue. If it really happened the way she had begun to tell it, why did she not say so in the letter written only a few days after the event? Geza argued that Carole had inadvertently created a false memory of the moment, the supposed moment, when she and Ferdinand came to the high edge of the Kahama falls and were momentarily tricked by a vegetative trompe l'oeil. Geza also insisted that Ruth's geological education and her intense interest in the subject should have kept her from failing to see the edge of the falls. "Every time she went some place, she was aware of what she was looking at," he insisted, and that awareness would imply that Ruth knew where she was and where the cliffs and waterfall could predictably be found as she followed Kahama Stream down toward the lake on her way back to camp.

But if she didn't jump deliberately or fall accidentally, what does that leave? The third way to fall—by being pushed—is an alternative that Geza was reluctant to consider in full. During our extended conversations, I observed the painful ebb and flow of this third theory as he struggled with an uncertainty about the facts combined with the normal disbelief most ordinary people would have about such a radical alternative. The idea was never fixed or consistent. He "wavered," he told me—and yet still, as he once summarized the basic problem: "Something is not right about this whole business, and I don't know what it is."

· · ·

The *something not right* Geza referred to was complicated, and for me to grasp it required, first of all, reviewing the hundreds of letters, emails, documents, stories, and assessments he had gathered into those five very large binders. In the end, I began to understand that it was important to consider not what was available and placed before me in those five binders, but what was missing—and to think about how the presence of absence, that state of apparent *missingness*, might have happened.

As I mentioned earlier, when Geza returned to Gombe in the summer of 1970, he discovered a strange scatter of things-not-there that should have been. Ruth's old hut, which had previously been his, was completely cleared out and being inhabited by a new person. And so, where were the missing items of his own that had been stored in the rondavel?

Years later, Geza acquired a copy of a journal Ruth's mother, Dorothy Davis, had written when she and her husband, Price, traveled to Gombe in 1969 for their daughter's funeral. Following a long flight in a small plane from Nairobi, Hugo van Lawick arrived with Dorothy and Price Davis at around seven thirty in the evening of July 23. Ruth's parents were exhausted and despondent. They rested briefly at the beach hut—the Cage—and, after a quick dinner, turned in for the night. After breakfast the next day, they endured the necessary marathon of greeting a long line of mourners, followed by the memorial service and burial, followed by another necessary marathon of farewells until it was time to consume dinner and be consumed by sleep. Not until after lunch the next day, July 25, were the Davises brought upstairs and shown Ruth's hut. What they found must have upset them. Mrs. Davis wrote, "I had expected to find it just as she had left it and that her presence would be very much there." Instead, they found the place "bare," since "the girls," meaning Lori Baldwin and Cathy Clark, "had packed everything." With what seems to me like a forced graciousness, Mrs. Davis finally concluded that the cleaning and packing up of everything in Ruth's rondavel was actually an example of "beautiful thoughtfulness and . . . help over a difficult place." She and her husband sat on the floor and sorted through what was there, and they did so decisively. "Most everything we just left and asked the researchers to dispose of clothing and such personal items as we decided not to bring home with us." They then packed a single suitcase in order "to take with us [some] of her books and such personal things as we wished to keep."

The mother's account never describes which personal things she and her husband took with them and which they left. It is reasonable to believe that Geza's own possessions stored in the hut—clothes and camera equipment, for example—were given away then and thus are permanently gone. But Ruth had saved Geza's letters, and they, too, never again surfaced. Where are they? When Ruth's mother wrote to Geza, after she and her husband returned to America, she mentioned nothing about his letters or what they had taken of Ruth's possessions. Geza later imagined that they would have taken some of their daughter's favorite books and other personal mementoes, but Ruth's own personal journal should have been by far the most compelling thing, and yet the parents never mentioned it.

The journal was impossible to overlook. It was a highly visible object, clearly an important document. While Geza was there, she typed her entries regularly, most evenings. By the time she disappeared on July 12, her journal should have contained several hundred typewritten, single-spaced, legal-size pages. The pages were punched with two holes and fixed into an office-style document binder, which had a pink cover with the trade name, Kant's Flat File, embossed on the front. Geza, who spent many hours watching or listening to Ruth type up those entries knew that she always kept the binder in her hut and out in the open. So, it is reasonable to ask: What happened to Ruth's personal journal?

After the search for Ruth was over, Nic Pickford prepared an official report in multiple copies for each of the six days spent looking for Ruth, describing where each search party went, and he included with that report several supplementary maps. The original report and maps and all copies of them seem to have disappeared. What happened to the official report and maps?

Then there was the B-record. Geza was distressed back in 1970 as he looked through the filing cabinets at Pan Palace to find that the B-record maps were missing. It was part of the established procedure that each B-record was accompanied by a hand-drawn route map showing in detail what route the target chimpanzee had taken that day. What happened to the B-record maps?

Thirty-seven years later, Geza was able to consider the B-record at a more leisurely pace. By then, all the Gombe records, starting from the first field notes written back in July 1960, had been consolidated and archived and were being digitized for computer access by a former Gombe student from the early 1970s, Anne Pusey. Pusey began this massive project in the 1980s while she was a professor at the University of Minnesota; Geza communicated with her about Ruth and the B-record between 2006 and 2008. But when Geza began to review an electronic version of this massive archive, the existing B-record sent by Professor Pusey, he found anomalous gaps in the very places where he had hoped to learn something about Ruth's movements during the weeks before her death. What happened to those missing pieces of B-record?

In short, several things had disappeared in what seemed like a deliberate pattern of selective amnesia. Almost nothing about Ruth's final weeks remained. And in his darker moments, Geza began to imagine that that pattern was, in its totality, a faint echo of the noise produced by a hidden and malicious actor, someone who tried to conceal the location of Ruth's body from the searchers.

. . .

A logical flaw to that dark theory, it eventually became clear, is chronological. At least one of the important missing documents—Nic's official report—was obviously created after the body was found, so it could not have been disappeared in order to obscure the location of the body. Given that chronological clue, along with some others, Geza finally came to believe that the person who most likely created the pattern of amnesia was not a malicious actor but an anxious one named Hugo van Lawick.

Hugo was there at the right time. His stay at Gombe that July was defined and chronologically framed by his role as official escort for the grieving Mr. and Mrs. Davis during their small-plane flight from Nairobi to Kigoma on the evening of July 23 and back again to Nairobi on the morning of July 26. He left after Nic had completed his official report of the search, and he arrived a couple of days before the Davises were shown Ruth's hut and failed, apparently, to find her personal journal.

Hugo had been in Europe with Jane and Grub, attending and celebrating his brother's wedding, when Ruth disappeared, and so his motivation could not possibly have been to affect the direction or quality of the search. His motivation might have been simpler and more mundane than that. Hugo could have been worried about potential lawsuits directed against the Gombe project and perhaps against himself and his family. It would be natural to worry about such a possibility under those circumstances, and it could be that Hugo's concerns were amplified by conversations with his brother, Godi, who was an attorney in Holland. Thus, Hugo might have had a motive for tampering with and destroying documents: to erase anything having to do with Ruth and her emotional condition during the last few weeks of her life, including her self-described "depression" and her general state of unhappiness.

Hugo also had the opportunity. Indeed, as codirector of the Gombe project, he had the extraordinary opportunity to go anywhere in camp and do almost anything he liked at any time without having to justify his actions to anyone. After Hugo arrived with Mr. and Mrs. Davis at Gombe on the evening of July 23, they went to bed as soon as they were able. He could have spent hours scouring Ruth's hut and going through Ruth's personal items at that time. Cathy and Lori sorted and packed up Ruth's things at the direction of Steve Stevenson, which must have happened before Hugo arrived. But Hugo's search for any personal items of Ruth's that seemed potentially damaging to the operation would then have been greatly simplified. Hugo similarly had the opportunity to go through filing cabinets in Pan Palace at his own pace and to remove all of Ruth's hand-drawn maps in the B-record as well as any of the entries he imagined might be of concern.

Everyone was under stress during the funeral, and Hugo must not have been pleased about being forced to cut short his family time in Holland. It's clear that he and Jane were by then having serious financial problems; I believe they were also by then experiencing marital conflicts that would soon lead to their estrangement, separation, and then divorce. So Hugo may have felt upset and under pressure when he showed up at Gombe for the funeral, but was he actually capable of removing and destroying documents from the files and personal items from Ruth's hut? Geza thought he was.

. . .

In the course of working to reconstruct the memory of that long-ago time, Geza managed to get in touch with most of the people who had been at Gombe during the latter half of 1968 and the first half of 1969. He sought out contact information for Nic Pickford somewhat late in this process; but by early March 2007, he had discovered a telephone number for Nic, who was living in semiretirement in Southern California.

Geza dialed the number. After the second ring, he heard a voice with a trace of Kenyan-British accent. The voice was different from what he remembered, rougher perhaps, but marked still by familiar inflections and phrasing. Geza asked: Was he born in Kenya? Had he worked at Gombe? After an affirmative response to both questions, Geza identified himself, and Nic responded warmly: "I remember you well! I remember you well!"

Nic was happy to hear from Geza, and Geza was moved by the enthusiastic response and by Nic's unhesitating openness about recalling the past. He seemed just as Geza remembered him: calm, helpful, straightforward. Since he had been camp manager, rather than a researcher, he was in a position to know more than most about practical developments and the comings and goings of people. He had worked closely with Hugo and had directed the search for Ruth's body, and he also happened to have an excellent memory. Nic's memory of Hugo, during the time he was there with Mr. and Mrs. Davis for the funeral, was not of a concerned or patient or grieving man but rather of an impatient and angry one. It seemed to Nic as if Hugo was angry at the researchers, as if somehow Ruth's death and disappearance had been their fault.

Nic and Geza continued their catching-up communications: phone calls and emails followed by, in September of that year, Nic's visit to Bethesda, Maryland, where he stayed with Geza and Heather for a few days. That extended contact with Nic added another dimension to the story and to Geza's thinking by, first of all, reminding him of what an impressive trek Ruth must have made on the afternoon of July 12, 1969. She had gone far beyond what anyone in camp had imagined. It was a long distance in a short time; and since both Nic and Geza understood through intimate experience the challenging nature of Gombe's terrain, they recognized that the satellite's-eye distance from camp to the Kahama Valley falls, which was perhaps two miles, would have in reality amounted to about six to eight miles involving a vertical gain in elevation of somewhere between fifteen hundred and two thousand feet. Ruth covered that distance following chimpanzees over a period of three to three and a half hours on a hot afternoon.

She weighed about 105 pounds. She had been missing dinners, which meant she would have been snacking on leftovers in Pan Palace. In other words, she had some serious limitations even before she started, and the follow itself must have been exhausting. Such was confirmed by Carole's memory of the tape, which suggests that Ruth was breathing hard, gasping between words, speaking in broken sentences. When Ruth lost the chimps, according to Carole's reconstruction, she was still comparatively high in the valley, still some distance away from the waterfall. At that point, she would have had two choices. She could have climbed up and found the trail on the ridge—the Mkenke Watu path—or she could have simply followed the stream down through the valley. Because she seldom carried water, Ruth would typically look for a stream to quench her thirst. It made sense, then, that she did seek and find the stream, and from there headed downhill. It was not yet dark. The tape recording had stopped at around 3:45, and she would have started not long after that, albeit in a state of serious exhaustion, to follow the stream.

Aside from an appreciation of the physical stresses of that day's follow, Nic added, in a later email, another important piece of information. After all the work done in Kigoma following the discovery of Ruth's body—visiting the morgue, reporting to the police, and so on—Nic returned to Gombe and walked up Kahama Stream in order to consider freshly the place where Ruth's body had been found. He never climbed up to the top of the falls. He stood at the base, at the edge of the pool, and he looked up, carefully examining the cliff face. He noticed something anomalous in the vegetation to one side of the falls: "I remember looking carefully at the cliff face and seeing the path she had followed as she fell, there were bushes bent down and a scar (fresh earth) in the red dirt at the top where the fall had started," Nic wrote.

The "path" Nic referred to was something typical for an obstacle like a falls or high cliff at Gombe. It was path made by animals: bush pigs, perhaps, or chimpanzees. "The path was at the absolute edge of the cliff face," according to Nic, "and although the chimps would have had no trouble walking it, a human would have had to bend low to avoid bushes and would have had to grasp bushes to help stay on the path." Nic's conclusion: Ruth grabbed a bush and used it to swing around. The bush gave way, and she dropped headfirst over the cliff and onto the rocks below.

5.

That is a reasonable and reassuring conclusion, one that Geza came to accept. I do not. At least not completely. I don't believe that Ruth fell simply as the result of an unfortunate accident. Nor do I believe that she intentionally jumped. Nor do I imagine that she was deliberately pushed. None of these three ways to fall explains fully or clearly what happened.

The problem is our natural tendency to think in categorical terms about shattering events, such as the sudden death of someone young and promising, as if they represent intellectual puzzles to solve in the style of a fictional detective. Unlike an artfully packaged mystery in a work of fiction, this unpackaged mystery from long-ago real life will never be so neatly resolved. Many things are gone. Important people have died. The world has shifted. More to the point, however, we have so far been examining this drama without fully considering its fundamental nature, which is psychological rather than physical.

Let me be more specific. When the searchers first went looking for Ruth, they imagined precipitating events based on the most obvious dangers of a place like Gombe: bad case of malaria, serious fall, bite of a poisonous snake. They failed to consider an additional danger, which is the possibility of becoming emotionally disconnected from one's human companions to the point of severe social alienation.

I don't believe anyone intentionally shunned or ostracized Ruth. In fact, I can believe quite the opposite, that the others in camp made sincere efforts to include her within their small community. Those efforts were not successful. Ruth was reluctant to talk in the chattering style, exceptionally slow to reveal herself or make new friends. She was a classic introvert. After Geza left, she found herself in a situation of cultural and social tension where her somewhat passive introversion was challenged by an assertive extroversion, and she found her own strong beliefs about how to do animal behavior science seemingly under attack from others who had learned a different kind of animal behavior science.

Ultimately, though, the social isolation that Ruth experienced at Gombe was ordinary rather than extraordinary. It happens in a less intense form to people everywhere, because people are members of a group-living species. The glue of belonging that holds together groups—of people, of chimpanzees, or of any of the thousands of other social species within the mammalian group—is an emotional and psychological substance that consists of contrary, push-and-pull affective systems that give, on the one hand, that painful stab of social rejection and, on the other, the liberating pleasure of social gregariousness. These are such universal emotional or affective experiences that we often stop recognizing them consciously, and so we unconsciously arrange our lives—through careers and marriages, stable families, steady friends—in a way that achieves a pleasing stasis, a steady sense of intimate belonging. This occurs whether we think about it or not, and usually we do not.

The need to belong is not a lust of the flesh but a hunger of the soul, and since the experience of belonging becomes lodged in memory, it exists as an essential part of the self. *Not* belonging, in its extreme version—that is, severe social alienation—can be painful indeed, and ultimately it will have damaging effects. Describing the experience as "loneliness" entirely misses the point. Describing it as "depression" fails to clarify the cause. And while depression most often makes people feel sad, severe alienation can make them feel mad. A failure to belong has the potential of turning people into angry misanthropes or even violent terrorists.

In explaining the root of Ruth's mental or emotional difficulties, Geza always preferred the theory of a sudden hormonal imbalance caused by discontinuing her birth control pills. While I can imagine that a sudden change in her hormonal economy could have had a significant effect on her mood and mind, I consider it secondary to much more potent events and situations. Learning in mid-May that her best friend from high school had been killed in Vietnam must have been devastating. And surely even more devastating was the realization that there was no one in camp, no intimate friend or quiet ally, with whom she felt she could talk about it. Even worse than her grief, then, was her sense of isolation.

By July 4, after "long periods of depression," she was attempting to "straighten myself out," as she phrased it, which involved "going deeper into my shell." She had found "an endless number of things here, both big and small, that I could concern myself with and worry about, but I have decided not to. The only thing I care about is the welfare of the chimps and when that is in danger, I will fight until death (literally!). I do not understand the people around me and, since I have no respect for most of them, I cannot be bothered trying." Those words were written eight days before she died, and one is struck by the phrase: *I will fight until death (literally!).* It is a strikingly self-conscious pronouncement. But there is no evidence that Ruth's declaration was anything beyond dramatic hyperbole declared in anger. Ruth wrote other angry things, of course, but it is clear to me, finally, that a more revealing sentence is the one immediately following: *I do not understand the people around me and, since I have no respect for most of them, I cannot be bothered trying.*

. . .

Ruth's failure to find an intimate connection within that tiny human community was compensated for in part by her success in knowing the chimps and her sense of discovering an intimate belonging with them.

She was fortunate to have that experience. Gombe in the late 1960s was the one place on earth where a few chimpanzees had accepted a few humans as part of their world without being forced to, without the usual case of people locking them up in prison in order to learn more, and it was run by someone who allowed people doing research to develop in more than one direction. Jane Goodall let people follow their interests, and she did not insist on everyone using the same protocol. If Ruth had gone on to do academic work at some venerable institution, she would have been expected to frame her research in a certain way. She would probably have been forced to learn theory and statistics and analytical methods, and then, after a few years, she might have become a supposedly objective observer within a professional culture that identified certain particular systems as appropriate for doing, seeing, and representing. At Gombe, however, she was starting to know the chimps directly and intuitively and emotionally, becoming in her own skilled amateur's way a watchful presence. In spite of the obvious dangers in doing that, she was bravely moving farther into the chimps' physical and social worlds, for longer periods of time, than anyone else was doing or, conceivably—with the exception of Jane Goodall herself—had done.

Among people, Ruth was a natural outsider: a fish out of water, an odd duck. Such dismissive clichés are sometimes used by people who are natural insiders. These are the people who find it simple to blend in, whose lives are made easier by a quick facility with the appropriate phrase, the engaging smile, the right gesture and facial expression at the right moment. Natural insiders are fish in water and ordinary ducks, and their absence of social uncertainty or questioning can also make them predictable and even staid creatures. They see what others see, think as others think, believe what others believe, are comforted most when dressed in a social camouflage that disguises the world's jagged edges and inconvenient truths. Outsiders, because of their distinctive social weaknesses, can be colossal failures. Even among chimpanzees this may be true. Worzle, conceivably because of his strangely human-oid eyes, was low in the dominance hierarchy, and perhaps we can say he suffered as an outsider. But those same weaknesses combined with a certain degree of intelligence, craft, and intensity of will can also make outsiders successful, especially given their capacity to see what others don't see and to think what others don't think. Perhaps Mike, with his clever wizardry, his use of stolen kerosene tins to astonish and frighten the others, is an example of the successful outsider. But among humans at least, an outsider with the right ability and training might hope to find real success as a creative artist or a creative scientist or simply a creative person.

Someone like Jane Goodall, whose talents and self-confidence combined with an outsider's remarkable vision to help revolutionize the science of animal watching in the middle of the twentieth century.

Or someone like Alphonse, who made his place on the shores of Gombe as a flawed man unable to stand upright in his pirogue because he had only one foot, a lost Mfipa from the south come north to fish alongside the Waha—and, finally, a creative fisherman who introduced the long drift net, the makira, to that part of the world. It was always true that the makira might become unattached from its moorings and drift away. In the dry season especially, an early morning wind could appear, blowing strongly from the east, and that strong wind might become stronger and push an unattached makira out toward the middle of the lake, where the waves are huge and fierce and the net could be lost forever. That far out on the lake, a person could become lost forever, too, which was Alphonse's fate a few years after Ruth fell off the cliff. Alphonse disappeared one morning, having followed his drifting makira too far out on the lake, and it might be said that Ruth, too, disappeared while following her net too far.

She was not deliberately pushed by anyone, but she was pushed by social circumstances. She did not intentionally jump, but, troubled and even desperate, she pushed herself too far for too long, so when she came to the edge of the Kahama falls she arrived without her full capacity for observation and planning. She fell accidentally, but only because, conceivably, she followed the chimpanzee path that went around the top of the falls and then, at a critical moment, tried to make a move, grab a bush perhaps or do something else that would be hard for a weak, two-handed and two-footed human ape but easy for a powerful four-handed chimpanzee ape.

And so she fell.

Acknowledgments

I am grateful to those people who lived and worked at Gombe during the late 1960s and contributed to this book with interviews and commentary, particularly Jane Goodall, Geza Teleki, Carole Gale, Nicolas Pickford, Timothy Clutton-Brock, Patrick McGinnis, and Timothy Ransom. Tim Ransom's interviews have been supplemented with reference to material from his book *Beach Troop of the Gombe* (1981), while Tim Clutton-Brock's contribution was clarified by reference to his 1974 *Nature* article, "Primate Social Organisation and Ecology," and his 1975 piece in *Folia Primatologica,* "Feeding Behaviour of Red Colobus and Black and White Colobus in East Africa." A few others who arrived not very long after Ruth's death and spent significant time there—including David Bygott, Richard Wrangham, and Anne Pusey—also contributed with interviews, information, and opinions, and I thank them for their generous help. Anne Pusey is currently the J. B. Duke Professor of Evolutionary Anthropology at Duke University and director of the Jane Goodall Research Center at Duke, which today holds the archives of long-term data from Gombe. These include records and observations originally written by Ruth, Geza, and Carole in 1968 and 1969. Geza and I spent three days at Duke, and Anne generously spoke with us about the Gombe record-keeping methods and enabled us to dip as deeply as we wished into the archives and review relevant records.

To supplement the many dozens of hours of interviews conducted specifically for this book, I have relied upon journals, letters, informal written recollections, and official documents and records. Additional materials include five large files of written recollections, correspondence, and photographs provided by Geza, personal journals and letters from Carole, and an extraordinary personal correspondence lent me by Jane Goodall. The Goodall letters were, in fact, gathered, edited, and published on their own during the ten years that I did the research for her biography, *Jane Goodall: The Woman Who Redefined Men* (2006). And since *The Ghosts of Gombe* is, in some ways, an expansion of that earlier work, albeit with a very different style and orientation, I should mention that a significant amount of the background information for this book comes from the biography and, in turn, from the gracious contributions made by dozens of people named in the acknowledgments there. I am grateful once more to those many dozens.

I have stayed at Gombe twice, and thus my knowledge of the place is somewhat anchored in direct experience, although inevitably that of a social and cultural outsider. But I have also, in this book, tried to expand my vision of Gombe by suggesting some of its cultural and historical context beyond and before the Euro-American one. Mrisho Mpambije Kagoha, a Lake Tanganyika fisherman who once fished off the shores of Gombe, and who met Jane when she first arrived in 1960, gave me his impressions of that interesting encounter during an interview conducted in the mid-1990s, with the translations between Swahili and English provided by David Anthony Collins. Tony Collins, as I've long known him, came from England, soon after Ruth died, as a graduate student researcher who intended to study the baboons at Gombe. He stayed and remains there today as an important member of the continuing community and also as director of Gombe's ongoing baboon research. Tony is himself an astute cultural historian for that part of the world and has kindly served as my informant—aided significantly in that role by Shadrack Mkolle Kamenya (director of chimpanzee research and conservation with the Jane Goodall Institute, Tanzania), who grew up in a village at the edge of Lake Tanganyika. Other important Tanzanian-born informants long associated with Gombe and contacted through Tony Collins include Jumanne Kikwale, Hamisi Matama, and Eslom Mpongo. Smita Dharsi, yet another local informant, was a few months old when, in 1960, she met Jane in Kigoma, which, as the town nearest to the Gombe research station, was and still is its main source of food and supplies. The name Dharsi should be familiar to readers of this book; Smita's father, mother, and grandfather, who were close friends of Jane's during the 1960s, ran a general store and petrol station in town that served the Gombe research community in both practical and personal ways.

Further information arrived through contacts I made during 2013 and 2014 as a fellow at the Radcliffe Institute for Advanced Study at Harvard University. My Radcliffe researchers Matthew Wozny and Deborah Onuoha helped expand my knowledge of East African cultures and beliefs, while Harvard undergraduate Kennedy Mmasi, a Tanzanian by birth, added his own distinctive recollections and perspectives. In addition, I am indebted particularly to the following three written sources: Kefa M. Otiso's *Culture and Customs of Tanzania* (2013); Gary Van Wyk's *Shangaa: Art of Tanzania* (2013); and Michele Wagner's "'Nature in the Mind' in Nineteenth- and Early Twentieth-Century Buha, Western Tanzania," which is a chapter in *Custodians of the Land,* edited by Maddox, Giblin, and Kimambo (1996).

In 2015, I was given an opportunity to consider the psychological dimensions of this story while a visiting scholar at the Erikson Institute for Research and Education, part of a large and advanced psychiatric treatment center in western Massachusetts called the Austen Riggs Center. There, I was settled into an office on the periphery of an excellent library and surrounded by professionals who have dedicated themselves to psychotherapy in an institutional setting. My time as an Erikson Scholar enabled me to examine this story as both a tale of someone who died and a tale of those who survived—how they were affected emotionally and sometimes damaged by that experience. I must thank John Muller and Virginia Demos, who served as teachers and mentors while I was at Austen Riggs; I also want to express my gratitude for additional support from Jane Tillman and Lee Watroba.

Finally, I must thank three special people who served as friends, advisors, and skilled readers of the manuscript during its several stages of revision: Heather McGiffin, Wyn Kelley, and Daniel C. Dennett.

Sober nonfiction frequently presents a single perspective, the author's, on some purportedly truthful account. I have worked to transcend that kind of sobriety in large part by recognizing openly that the past can never be perfectly captured and by presenting multiple and sometimes conflicting perspectives. I express the perspectives often, though not always, of those individuals who consented to interviews, and I try to identify them sometimes directly ("in her opinion," "it seemed to him") and sometimes through a close paraphrase that echoes the individual's speaking style. In instances where I represent someone's actual thoughts by placing the words in italics, I am quoting directly from something described as a thought in an interview or from something written privately (as in a journal). When I refer to language written publicly (as in letters or published material), I place it in quote marks. I also use quote marks to identify conversations that were written down or remembered as direct dialogue while recognizing, of course, that people commonly maintain inexact recollections of conversations that took place in the distant past. I have in one instance changed a name out of respect for someone's privacy; but probably my closest approach to fiction occurs in the brief portrait of Alphonse the fisherman, which has been developed from photographs, background research, and other people's recollections of the man. In spite of such creative liberties taken, the story told here is true—or as true as I can make it in shape and fact.

Dramatis Personae

(November 1967 to July 1969)

MANAGEMENT AND STAFF

JANE VAN LAWICK-GOODALL, British primatologist and codirector of the research center

HUGO VAN LAWICK, Dutch photographer and codirector of the center

NICOLAS PICKFORD, Kenyan-British camp manager from June 1968

MARGARET PICKFORD, South African bookkeeper, June 1968 to April 1969

RASHIDI KIKWALE, the *mzee* (honored elder) and head of staff

HILALI MATAMA, hired in 1968 as first field assistant

Other field staff hired by the summer of 1969: Eslom Mpongo, Hamisi Mkono, Jumanne Mkukwe, and Yusufu Mvruganyi

DOMINIC BANDORA, cook

SADIKI RUKUMATA, assistant cook

MPOFU, carpenter and boatman

CHIMPANZEE RESEARCHERS

RUTH DAVIS, American long-term volunteer, arrived in May 1968, died on July 12, 1969

CAROLE GALE, American long-term volunteer from the Friends World Institute in Nairobi, arrived in November 1967

GEZA TELEKI, American graduate student from Pennsylvania State University, February 1968 to March 1969

PATRICK MCGINNIS, senior researcher; on temporary leave from Gombe, January 1969 to early July 1969

PATRICIA MOEHLMAN, senior researcher, finished near the end of November 1967

ALICE SOREM, senior researcher, finished in November 1968

LORETTA BALDWIN, American long-term volunteer, arrived in March 1969

CATHERINE CLARK, American long-term volunteer, arrived in November 1968

OTHER RESEARCHERS

TIM RANSOM, American graduate student and baboon researcher, finished in May 1969

BONNIE RANSOM, American graduate student and baboon researcher, arrived in January 1968, left in December 1968

TIMOTHY CLUTTON-BROCK, British graduate student and red colobus monkey researcher, arrived in January 1969

MICHAEL SIMPSON, British senior scientist, arrived in January 1969

TANZANIAN PARKS RANGER

FERDINAND UMPONO, first parks ranger to live at the new tourist station in Nyasanga Valley

THE FISHERMEN

IDDI MATATA, unofficial spokesman for the *dagaa* fishermen

ALPHONSE, fisherman from the south who used a drift net (*makira*)

VISITING DISTINGUISHED SCIENTISTS

DAVID HAMBURG, psychiatry professor at Stanford University

ROBERT HINDE, ethologist and Jane's former advisor from Cambridge University

PHYLLIS JAY, primatologist and anthropologist from the University of California, Berkeley

CHIMPANZEES

High-Ranking Adult Males

CHARLIE, Hugh's predictable ally and probably his younger brother, bold and powerfully built

HUGH, probably Charlie's older brother; Ruth's favorite

HUMPHREY, very big and aggressive, known for his attacks against chimp females; he also seemed to dislike human females

MIKE, a comparatively benign alpha who acquired his top status in 1964 through the creative use of stolen kerosene tins

The Flo Family

FLO, the old matriarch: aggressive, high-ranking, and reproductively successful

FABEN, Flo's teenage son and Figan's older brother, crippled by polio in the 1966 epidemic; one of Carole's favorites

FIGAN, Flo's teenage son and Faben's ambitious younger brother; another of Carole's favorites

FIFI, Flo's preadolescent daughter

FLINT, Flo's juvenile son

FLAME, Flo's infant, born in August 1968, disappeared (apparently died) in January 1969

Other Researchers' Favorites

DAVID GREYBEARD, Jane Goodall's early favorite, disappeared during a epidemic of respiratory infections in the late spring of 1968

LEAKEY, an older, middle-ranking male, who could have been a brother to Worzle; Leakey was also Geza's "friend" and introduced him to termite fishing

Others

DÉ, young adult male

EVERED, elder son of Olly, ambitious but lacking close allies

GOBLIN, born to Melissa in 1964

GODI, adolescent male

GOLIATH, powerful male and former alpha past his prime

MELISSA, high-ranking female

OLLY, shy, timid female, died in 1969

PALLAS, adult female

PEPE, male crippled by polio, died in 1968

POOCH, young female with a serious injury, died in 1968

RIX, adult male probably killed in a fall from a tree, November 1968

SNIFF, young adult male

WILLY-WALLY, young adult male.

WORZLE, low-ranking older adult male marked by humanoid eyes (white sclera)

Illustrations and Credits

MAPS

Index

Milton Keynes UK
Ingram Content Group UK Ltd.
UKHW030626040224
437177UK00004B/63/J